HISTORY OF

FACTORY AND MINE HYGIENE

HISTORY OF
Factory and Mine Hygiene

Ludwig Teleky, M.D.

Columbia University Press

NEW YORK : MORNINGSIDE HEIGHTS

1948

COPYRIGHT 1948 COLUMBIA UNIVERSITY PRESS, NEW YORK

PUBLISHED IN GREAT BRITAIN AND INDIA
BY GEOFFREY CUMBERLEGE, OXFORD UNIVERSITY PRESS
LONDON AND BOMBAY

MANUFACTURED IN THE UNITED STATES OF AMERICA

To My Wife

Foreword

In 1912 there was held in Brussels a meeting of the International Congress on Occupational Accidents and Diseases. At one of the sessions devoted to diseases, a question was asked about measures in force in the United States to control industrial lead poisoning. The answer was given by the Chief Inspector of Factories of Belgium, Dr. Glibert, "But, it is well known that there is no industrial hygiene in the United States. Ça n'existe pas." Fortunately one of the three Americans who listened to that damning statement was Charles Neil, Commissioner of the newly established Bureau of Labor. Startled and mortified, he resolved to find out at least if it was true, and thus was initiated the first Federal inquiry into the vast subject of sickness caused by occupations.

Dr. Teleky's book gives us the history of this branch of medicine, which is also a branch of social science, from its earliest beginnings in antiquity down to its recent intensive development during the Second World War. It is fascinating reading, that early history. Take mercury, quicksilver, a poison which affects the mouth, with stomatitis and loss of teeth, and the nervous system, with shaking palsy and characteristic psychosis. There are great mercury mines in Almaden, Spain, which were worked in Roman days and are still worked, but so great has always been the danger of mercurial poisoning that under the Romans only slaves were used in mining the ore. Pliny the Elder, in writing of the diseases of slaves, mentions mercurialism, plumbism, and the consumption of potters and grinders. Later on, convicts were put to work in the Almaden mines. Justinian wrote that a sentence to these mines was almost equal to a death sentence. When finally free labor was introduced it was found necessary to limit the hours of work or too many miners would be incapacitated; therefore a tradition grew up, which was still in force in the twenties and may be now, that thirty-four hours, divided into eight days, should constitute a month's work.

The history of the efforts of physicians since the days of Pliny to protect workers against the dangers of their work is a history of very small beginnings, under men inspired by an unusual degree of humanitarian feeling or of scientific curiosity, and of rapid growth under the changed conditions brought about by the industrial revolution. Mining, not coal which came much later but metals, lead, copper, tin, sulphur, gold and silver, was what interested especially these early scientists. Some Germans have given us detailed descriptions of the work of metal miners in the sixteenth century and of "miners' phthisis," "miners' asthma," sudden death from poisonous fumes, and so on. Yet the German miner in those days had certain advantages which he lost later. In the early Middle Ages, in Germany as in all European countries, the doctrine (which we moderns regard as extremely radical) was generally held that the minerals in the earth were the property of the State, not the landowner. The State meant the sovereign, who in the German states gave the right to mine to groups of skilled men, independent, enjoying many privileges. They even had a system of workman's compensation, and their working day was only seven hours, their week only five days. But as mining went deeper and expensive apparatus was needed, capitalists had to come in and gradually the miner lost much of his favorable position. The mining industry in Germany, however, has always been the subject of scientific study and of governmental responsibility.

In England, characteristically, the movement to control the fearful conditions in the life of the miners, metal and coal, although helped by physicians, was largely the work of humanitarians. Dr. Teleky's description of the life of the Scottish, English, and Welsh miners are appalling, and the history of the struggle to better their lot leaves one with little respect for the "captains of industry." However, that struggle was successful at last and Dr. Teleky shows us that in the study and the control of industrial accidents and diseases it is England, still, that leads the way.

The history of industrial hygiene in the nineteenth century is one of rapid progress in Europe, whether through the combined efforts of physicians and humanitarian legislators as in England or through the will of an enlightened autocrat like Bismarck.

Hours were limited, child labor progressively controlled, women's labor regulated and measures introduced for the prevention of accidents and for the improvement of general sanitation. Toward the last half of the century, the study of diseases caused by dusts and poisons began to make great strides in Germany and in Britain. Dr. Teleky tells us of the setback to studies of the dust diseases which followed Koch's discovery of the tubercle bacillus. For years thereafter the attention of pathologists was diverted to bacteriological investigations, and the character of the inhaled dust was largely ignored.

Coming down to modern times the book gives us a wealth of material on the methods followed in the industrial countries to collect information concerning damage to health caused by occupations and to correct the conditions which lead to such damage. This is not altogether pleasant reading for a citizen of the United States. It is true, and Dr. Teleky does full justice to the fact, that industrial hygiene in this country has made great strides during its short life since the First World War. There is a vast amount of scientific research now being carried on into the dangerous dusts and gases, into the effects of radioactivity, of excessive heat and humidity, of excessive exertion, of the influence of diet and of fatigue. The Public Health Service, many of the universities, some of the great manufacturing companies, are publishing the results of studies on animals and human beings which are quite equal to the best European work and in some fields (as in the measurement of air contamination) lead the way. Dr. Teleky's list of these contributions is most impressive. But it is in the field of practical hygiene, of the application of our knowledge to actual conditions in industry, that American industrial hygiene falls far behind that of Britain and of the Continental countries.

Take the coal mines. American mine operators are proud of their mechanization, of American ingenuity and modern methods, as compared with the backward ways of the old countries. And yet, though our mines are not nearly so deep as those of England, Germany, and Belgium, which makes the problem of ventilation and accident prevention far easier for us, our rate of fatal mining accidents is the highest of all. It is in this industry that the contrast between American ways and European is most striking.

In Britain and on the Continent, factory inspectors are educated and specially trained persons. In England the early regulations required that the inspector be "a complete gentleman," and though that requirement no longer exists I can testify from an extensive experience that it is still largely followed. A leading American factory inspector said at a public meeting in 1940, "No factory inspection division expects factory inspectors to be chemical engineers, chemists, ventilating engineers or physicians." But that is exactly what European factory inspectors are expected to be. Moreover, they are civil servants, protected in their tenure of office. In our country of forty-eight states, this highly important function is performed in some forty-eight different ways, varying from an excellent system to one deplorably inadequate. In one highly industrialized state there is no civil service at all; the factory inspectors are turned out with every change of politics.

Under such a system there is no wonder that in many of the chiefly agricultural states industrial hygiene should be either nonexistent or utterly inefficient. But one would think that Kansas has enough industry to give some attention to the protection of workers. Kansas is one of the two states (nonindustrial Arizona is the other) which have no legal regulations for the control of industrial hygiene.

I heard a mining inspector of Oklahoma (they elect their inspectors out there) say in a public speech: "When I go into a mine and find a lot of things wrong and the owner tells me that if he has to put in all those changes he will go bankrupt, I just say 'Well brother, forget it!'"

That there are many American employers who need no compulsion to make them do everything possible to protect their workpeople against accident and disease, I would be among the first to acknowledge. But we cannot take that fact as a solution of the problem. It is as if one argued against the necessity of traffic regulations because every prudent, sober, considerate adult drives carefully of his own account.

It is impossible not to agree with Dr. Teleky's summary. The task remaining is first of all to put into more widespread use all the measures discovered to restrict dangers, control noxious substances, and generally to improve industrial health. Legislation in England and Germany and the efficiency of factory inspection in those countries point the way. Numerous laws, exact and detailed rules and regulations, are necessary, as well as their enforcement by highly qualified factory inspectors. With this as a basis, it is important to have the cooperation of associations, such as the International Labor Office, the National Safety Council, and also of trade unions, scientists and physicians.

ALICE HAMILTON, M.D.

Hadlyme, Conn.
September, 1947

Author's Preface

THE READER will find here only an outline of the history of industrial hygiene. A comprehensive study would require studies in the archives of the United States and those of the highly industrialized countries of Europe—England and Germany, also France, Belgium, and Italy—studies which cannot be carried out in these postwar times. Therefore this summary is based on my own library, rich in German, Austrian, and English material, and the extensive libraries of New York, among them the New York Public Library, the libraries of the New York Academy of Medicine, of the British Information Service and of the Engineering Societies.

I shall attempt to make a survey of the history of factory hygiene, that is, of the hygiene of the plant and its installations as well as that of the work itself, especially of health safeguards in dangerous occupations, and also of the protection of miners.

I realize fully that some other aspects of working conditions, such as the protection of women and children, the regulation of working hours and wages, have always been of much greater importance for public health and labor protection, because they protect the greatest number of working people in the most effective way. However, being a physician and industrial hygienist, I wish to confine myself to my own field, the hygiene of work. I shall mention the other phases of labor protection only in so far as they will enable the reader to follow the course and understand the development of the history. Accident prevention will also be treated, but since it is rather the concern of engineers it will be dealt with less thoroughly and without technical details. This survey also is restricted to the most industrialized countries: England, Germany, and the United States. Other countries are mentioned occasionally, and especially the important publications of other countries, science being international, legislation national.

Even in this restricted field I can give only a history of legislation and pertinent scientific progress. In the most industrialized states of Europe and in several states of the United States, the practice of factory hygiene parallels its theoretical growth, never quite keeping pace with legislative measures. There are other states in which there is a decidedly wide gap between legislation and practice. It is no doubt simpler to publish regulations and sign international treaties dealing with labor protection than to complement the laws by an efficient system of factory inspection.

It is almost impossible to acquire a real knowledge of factory hygiene as it is actually applied in different countries. Only a limited number and type of factories are open to the researcher and therefore only a very superficial impression can be gained. The one practical way to become acquainted with the state of factory hygiene in any particular country is to work there as a factory and mine inspector. This I did in Germany, and I am therefore quite familiar with the conditions existing in that country. However, I feel that my inspection tours in England and France, and my activity in the United States, did not produce information specific enough for my purpose and consequently these experiences will not figure in my study.

I wish to thank the Emergency Committee in Aid of Displaced Foreign Medical Scientists (Secretary-Treasurer Dr. Ernst Boas) for giving me time and leisure to devote to this work.

LUDWIG TELEKY, M.D.

New York
December, 1947

Contents

	Foreword by Alice Hamilton, M.D.	vii
	Author's Preface	xiii
1.	**Development of Industrial Hygiene** Antiquity, 3; The Middle Ages, 4; The 18th Century, 11; The 19th Century, 15	3
2.	**Legislation in Europe and the United States** Great Britain, 22; Germany, 37; United States, 51	22
3.	**Organizations and Associations Cooperating in Industrial Hygiene** The Workers' Unions, 75; Employers' Associations, 86; International Associations, 87; National Associations, 90	75
4.	**Institutions in Factories for Promotion of Industrial Hygiene** Safety Committees and Works Committees, 94; Safety Directors and Engineers, 99; Factory Physicians, 100	94
5.	**Methods of Preventing Injuries to Health** The Elimination of Dangerous Substances, 113; The Exclusion of Especially Endangered Groups, 116; Work Turnover, 124; Shorter Working Hours, 124; Cooperation of the Worker, 126; Technical Devices and Their Control, 127	112
6.	**Prevention of Accidents**	147
7.	**Education, Propaganda, Safety First**	150
8.	**Research and Studies** Governmental Research, 153; Nongovernmental Research and Studies, 166	153
9.	**The Progress in Other Sciences and Its Influence on Industrial Hygiene** Toxicology, 180; Hygiene and Physiology, 182; Statistics, 182	180

	CONTENTS	
10.	Special Problems	194
	Dust Diseases, 194; Caisson and Tunnel Work, 208; Skin Diseases, 215	
11.	Mines and Miners	220
	Legislation, 220; The Dangers and Their Control, 232; Rescue Work, 252; Ankylostomiasis, 254; Results of Mine Hygiene, 255	
12.	The Effects of Industrial Hygiene	264
13.	Summary	280
	Bibliography	285
	Index	319

HISTORY OF
FACTORY AND MINE HYGIENE

I

Development of Industrial Hygiene

ANTIQUITY

EFFORTS to protect man against injury from the tools he uses are very old. "Arm-protecting plates,"[1] that is, plates protecting the inside of the left wrist against the recoil of the bow string, seem to have been used in Central Europe in the Paleolithic Age (earlier Stone Age). In the Neolithic era (later Stone Age), they were made of clay, stone or bone, quadrangular or oblong, often curved, with two, four, or six holes for lacing in place. Widely used also in the late Neolithic and the beginning of the Bronze period, these were later replaced with gloves. Plates of this kind were also found in a grave prior to 1420 B.C. In an Aramaic relief of 730 B.C., tubes can be seen on the three fingers of the archer's right hand with which he is grasping the bow string. A Persian relief shows a device for the protection of the thumb and the hand of archers.

Other ancient pictures show leather rings[2] beneath the water vessels that the women are carrying on their heads. Similar rings are still in use today!

There are pictures of fluteplayers wearing strips of tape ($\phi o \rho \beta \epsilon \iota a$) on their cheeks and lips to prevent an excessive puffing of the cheeks that would eventually result in the relaxation of the muscles. This seems to be the first measure taken to avoid an occupational disease. Glassblowers in our own time sometimes acquire such flaccid cheeks, but they do not use any protective devices.

In the *Odyssey* (24th canto), old Laertes wears working gloves and gaiters as protection against thorns.

During the whole of antiquity, in the Middle Ages and

---

[1] Max Ebert, *Reallexicon der Vorgeschichte* (Berlin, 1924).
[2] *Katalog*, Internationale Hygiene Ausstellung (Dresden, 1911), Historische Abteilung.

even at the beginning of modern times, allusions to the protection of working men are very scarce. The most frequently quoted reference to industrial hygiene of that time is Pliny's statement that men engaged in the production of red lead cover their faces with bladders as a protection against dust.[3] This report may be authentic but the device itself is wholly ineffective. The bladders are airtight and the air necessary for breathing can only enter at the edge. Naturally, dust enters with it. But bladders and glass masks are repeatedly recommended in the literature, sometimes with special reference to Pliny, as in the writings of Ramazzini and Athanasius Kircher.

Physical deformities and damages caused by certain trades are occasionally reported by several authors of ancient times, beginning with Hippocrates who tells us that metal workers breathe heavily and suffer constrictions in the region of the stomach and that tailors have sagging stomachs. Such remarks are also to be found in Plato, Aristotle, Virgil, Martial, and others. Lead colic is described often, but nearly always as a nonoccupational disease.[4]

THE MIDDLE AGES

As in many other respects, economical as well as scientific, the Middle Ages were a time of retrogression, especially in the earlier periods. With the growth of guilds as organizations of employers in particular trades, first in

---

[3] In the Latin editions XXXIII, vii treats minium and cinnabar, distinguishing clearly between them, and referring to bladders used by workers with minium. The English translators (ed. 1601 and 1635) not being toxicologists, did not make this differentiation and used both words interchangeably, although minium is "red lead" (Pb_3O_4) and cinnabar is vermilion (HgS); both are used as a red pigment. In the English translation the story of the bladder is reported among workers with "Vermilion," and this mistake has reappeared in some modern books. Red lead is poisonous, but vermilion is not.

[4] L. E. Goldwater, "From Hippocrates to Ramazzini: Early History on Industrial Medicine," *Annals of Medical History*, VIII (1936), 27.

Germany in the 12th century and later in England, regulations were made to protect the masters and the consumers. After a while certain provision was made for sick, disabled, and aged workers, but guild rules contained no mention of protective measures. However, certain stipulations that were made for economic purposes and in order to reduce competition were incidentally beneficial to the health of workers. For instance, some guilds would not accept apprentices younger than fourteen, eighteen, or even nineteen years. There were also limits set to the working day. In London, the cutler's guild did not permit work to start before 4 A.M. or end after 8 P.M. in summer, and before 6 A.M. or after 6 P.M. in winter.

Legislation protected the neighborhood from vapors and waste substances. For example, in France, butchers and slaughterers were not allowed to slaughter in cities and they were compelled to dispose of waste matter (1366). In 1486, potters were forbidden to have their workshops in the residential districts of cities because of the harmful effects of the fumes. But in no country were there rules for the protection of workers themselves.

Delmege gives a description of industries in medieval England: [5]

Many of them were carried on in badly lighted and poorly ventilated shops in unsanitary surroundings by workers who seldom washed and who took no precautions such as are used in our time. Medieval working hours were from 4 or 5 A.M. to 7 or 8 P.M. in summer; in winter, roughly from dawn to dusk. Half an hour was allowed for breakfast, another half hour at about 3 P.M. and an hour and a half at midday for dinner and siesta . . . There was, however, no night work, no work on Sundays or on the more important holidays and the work ceased at 4 on the eves of such holidays and on Saturdays. Altogether, there were about 280 full working days in the year.

[5] J. A. Delmege, *Towards National Health* (London, 1931), p. 67.

Lead mining was extensive and this metal was used a great deal in pottery. Fire gilding was a process generally used. Poisoning must have been quite common.[6]

The fact that there was no real industrial hygiene in the Middle Ages is easily explained by the low levels of medical and technical science. Medical science, at that stage, did not recognize all the damages caused by occupational hazards; technical science had not advanced sufficiently to prevent them.

The end of the 14th century in England [7] saw the beginning of production on a large scale, with the wool trade apparently the first to grow. A 15th century pamphlet voices the grievances of the working classes. Wages were low, half being paid in merchandise which was doubled in price. Laws banning the "truck system" (payment of wages not in money but in merchandise or in cheques valid only in certain stores owned by the employer) were passed in Colchester in 1411 and in Norwich in 1460, but they were not enforced until several decades later. Regulations were also issued restricting the use of machinery (Norwich, 1478).

In Germany, we find the start of industrialism and

[6] This last statement is certainly not correct so far as it applies to pottery. Since the most common lead ore, galena, is practically insoluble in the fluids of the body, lead poisoning in mines is rare. As long as ground lead ore was used for glazing—and that was the case in the small potteries in the Alps up to the beginning of the 20th century and certainly so in earlier times in other countries also—there were very infrequent instances of lead poisoning in potteries. The danger of lead poisoning became great only when minium was used. This was especially true where pottery was a domestic industry (Hungary), and in factories where, because of the subdivision of labor, some workers were permanently employed in the process of glazing. But this did not happen before the end of the 18th century. The danger increased even more where white lead was used, as in England (Staffordshire) after the middle of the 18th century. On the Continent the development of pottery into a ceramic industry began about a century later. In the beginning of the 20th century began the struggle against lead poisoning in potteries, especially in England, where the introduction of fritting solved the problem to a great extent.

[7] E. Lipson, *The Economic History of England* (London, 1937).

capitalism a few decades later in the mining enterprises of the Fugger family. Foremost of this family from 1510 to 1525 was Jacob II, who managed large mines in the Alps. However, this development went no farther in Germany, where it was stopped perhaps by the destruction and disorder resulting from the Thirty Years War (1618–1648). In England, on the other hand, the capitalist system progressed and gave rise to a working class. At the end of the 16th century most of the textile workers were wage earners.

MODERN TIMES

First Steps in the Hygiene of Workshops and Mines. The booklet *Von den giftigen besen Temppfen ann Reuchen der Metal* (On the Poisonous and Noxious Vapors and Fumes of Metals), written by Ulrich Ellenbog in 1437, but not published until 1524, is termed by its editors Koelsch and Zoepfl (Munich, 1927) "the first publication in the world's literature that deals specifically with industrial hygiene." It is a popular brochure intended for use by goldsmiths and other metal workers, in which they could learn "how to protect themselves against the effects of silver, mercury and lead fumes." For protection against coal fumes the author recommends open windows, the burning of incense, and the sprinkling of wine on the coals. He suggests that the harmful effects of the fumes could be avoided by working in the open air, by averting the face, by covering the mouth with a cloth and by taking various medicines.

That the goldsmiths were the first to be thus cautioned is understandable, since in those times they were endangered more than any other tradesmen by the poisonous fumes of mercury, lead, and carbon monoxide. But the first extensive industrial hygienic installations were to be found in mines and smelters. Here the dangers were greatest and their effects immediately apparent. Mining is only possible when adequate measures are taken against its possible dangers.

In his *Zwölf Büchern vom Berg- und Hüttenwesen* (Twelve Books on Mining and Smelting, 1556), Georg Agricola, physician of Joachimstal and Chemnitz, makes many suggestions to the miners for their personal protection. He advises them to wear high rawhide boots because of the water; gloves up to the elbow; and loose veils over the face as protection against dust, which is more harmful than humidity when it enters the windpipe or lungs and which is also injurious to the eyes. He also gives a very fine description and excellent pictures of the devices for the intake and outlet of air in the mines. Although the shafts which he saw were, as a rule, not more than twenty-two meters deep—in other countries they were much deeper—ventilation was found to be necessary. This ventilation was obtained by building several shafts and, under certain circumstances, lighting a fire in the outlet shaft to make the air in it rise. An artificial air supply was produced by: (1) catching the wind and deflecting it by simple mechanisms in the mine; (2) machines with big wings, driven by wind, water, or man power; (3) bellows driven by water, men, or animals. These devices could also be used to dispel so-called "heavy damps." Agricola reports further that intelligent and trained miners ignited piles of wood on Friday evening and did not enter the shafts or tunnel before Monday, when the poisonous gases were dispelled. Less care was taken for the exhaustion of vapors in smelting plants. The pictures show various kinds of stoves which, in most cases, have no exhaust pipe at all, sometimes one that is too short. Hoods are missing or very imperfect. One of the stoves has a big condensing chamber on top, through which the fumes pass to escape through a chimney into the open air. Agricola recommends the building of such dust chambers in order to catch the denser parts of the fumes.

DEVELOPMENT OF INDUSTRIAL HYGIENE 9

Two or more chimneys may enter one such chamber. A picture shows us a man working at a furnace with his mouth covered by a cloth connected with his jacket and head gear.

In England the development of mine ventilation did not come until later.[8] In the middle of the 16th century John Cains writes on poisonous gases in coal mines "the miner simply swung his jacket to and fro for the purpose of creating a current of air." In the 17th century methane explosions became frequent. Many attempts were made to improve ventilation, by building a second shaft, the "ventilation shaft," or a "bye pit" and placing a basket with live coals in the bye pit, as first reported by Plots in his *History of Staffordshire*, 1686. In metal-ore mines, wooden pipes were installed in the principal passages, and through them air was blown into the shaft by means of bellows. Between 1516 and 1688, 317 patents for ventilation were granted in England.

In Germany, a book by Theophrastus von Hohenheim (Paracelsus) entitled *Von der Bergsucht und andern Krankheiten der Bergleute* (Phthisis and Other Diseases of Miners) originated in about the same years as the work of Agricola. It was written 1531–1534, printed 1567. It is an entirely different work, dealing very little with hygiene. Paracelsus points out that, since mouth and nose are open, breathing brings with it all kinds of harmful effects. However, he does not mention any method of preventing the entrance of harmful substances into the respiratory system. He recommends various drugs effective against such substances, particularly Essentia tartari.

Some other German authors of the 17th century may also be mentioned: Martinius Pansa (*Concilium peripneumonicum* 1614), Ursinus (*De morbis metallicorum*, 1652), and Athanasius Kircher (*Mundus subterraneus*, 1665), who

[8] Rosen, *The History of Miners' Diseases* (New York, 1943).

wrote about miners and smelters. These writers were all influenced by Paracelsus and did not contribute any essentially new facts, nor did they open any new vistas on the subject—as far as I could learn from the abstracts accessible to me (Ackermann, Koelsch, Rosen).

Samuel Stockhausen, physician of the Lüneburg mines, disputes Paracelsus in his *Tractatus de lithargyri fumo noxio morbifico ejusque metallico frequentiori morbo vulgo die Hüttenkatze cum appendice de montano asthmate, metallicis familiari, vulgo die Bergsucht* (Treatise on the Noxious Fumes of Litharge, Diseases Caused by Them, and Miners' Asthma), Goslar, 1656. He writes: "Prophylaxis rather than remedies should be used so that fogs, fumes, vapors, and metal dust be avoided and the miners try to preserve their strength." Dust compels the workers to tie scarves around mouth and nose. This book gives a good description of the "Bergsucht" (miner's phthisis), which consists, as he says, in a persistent shortness of breath that in some cases ends in a phthisis. He gives a very clear clinical picture of what we call today "silicosis." No less impressive is his description of the lead colic of the smelters. Written in Latin, the book has a German appendix destined for the miners.[9]

In the transactions of the Royal Society of England, the first articles on occupational diseases were published from about 1670 on. They deal with the production of white lead, mirrors, etc. There are special articles on miners and gases in mines by Jessop (Philosophical Transactions, 1675), Rogers (*ibid.*, 1676, 1677), and Leigh (*Natural History of Lancashire*, 1700).

In those centuries other writings with a belletristic tendency in the description of trades appeared in Germany, among them Jost Ammans, *Eigentliche Beschreibung aller*

[9] Unfortunately, I was unable to get this book (just as I could not get Pansa Ursinus, Kircher, and others); quotations are from the reports by Ackermann and Koelsch.

DEVELOPMENT OF INDUSTRIAL HYGIENE

Stände mit Reimen von Hans Sachs (Descriptions of All Professions and Trades, with Rhymes by Hans Sachs), published in 1568 in Frankfurt, and *Christian Weigels Ständebuch* of the year 1698. Neither book contains in the text or in the numerous pictures anything concerning industrial hygiene, unless one applies this term to the picture of a baker standing before his stove and wearing nothing but an apron. Most of the tradesmen in the illustrations are very well dressed and seem to wear their holiday clothes rather than work clothes.

THE 18TH CENTURY

Ramazzini and His Successors. It is amazing that not more remarks about the hygiene of trades and workshops can be found in the famous book which was the first to treat of occupational diseases as a whole: Bernardo Ramazzini's *De morbis artificum diatriba* (Treatise on the Diseases of Artisans), Modena, 1700. It contains many directions for personal prophylaxis, such as bathing, changing of clothes, correct posture, physical exercises, wearing trusses, and covering the mouth with shawls in dusty trades. Rarely do we find any remarks about the hygiene of workrooms or tools such as the advice that subterranean grain stores should be well aired before workmen enter them, or that certain workrooms should be spacious. More frequent is the recommendation of prophylactic medicaments against damages caused by this work. The relatively small interest in workshop conditions is perhaps explained by the fact that the author as a physician was more interested in the clinical aspect of diseases, and he is to be highly praised for paying attention to the circumstances which caused them. But he has not attempted to enter the technological field, although this would be necessary in any attempt to deal with prophylaxis. But in describing the working conditions of the individual trades and the damages caused by them,

Ramazzini laid the foundation of industrial hygiene and the basis for prophylaxis.

The great importance of this work may be seen in the fact that it was reprinted six times before 1745, and that there were, according to Koelsch, in the 18th and in the beginning of the 19th century four Italian, three French, five German, and four English translations. Most of these were not mere translations. In some cases, the "translator" added his own experiences—we shall report some of them—while others published excerpts of Ramazzini's book, sometimes without mentioning the author (for example, Hecquet, 1740) and sometimes compressing its contents into a few pages, as did Nicolaus Skragge in Upsala in 1764 and Buchan in his *Medicine Domestique*, in 1775. Such "translations" were common throughout the following century.

Other publications of the 18th century dealt with ore mines and metallurgical plants which in addition to a few other industries at that time brought about specific dangers and damages to the worker's health. Such books are Friedrich Hoffmann, *De metallurgia morbifica*, 1705 (On Illnesses Caused by Metallurgic Work), Michael Alberti, *De praeferandis metallicorum morbis*, 1721 (Special Illnesses of Metal Workers), and Johann Friedrich Henckel, *Medizinischer Aufstand und Schmelzbogen, von der Bergsucht und Hüttenkatze*, 1745 (Medical Treatises on Miners' Phthisis and Smelters' Illnesses). The latter recommends dust detection, an improved form of which was used even 180 years later: "If you put a plain stone or a sheet of glass beside you, you will see with astonishment what a great amount of dust accumulates in 6–8 hours."

C. L. Scheffler's *Abhandlung von der Gesundheit der Bergleute*, 1770 (Treatise on Miners' Health) discusses lung diseases extensively but tries in vain to develop greater insight into them.

In 1716, Friedrich Hoffmann wrote another very important book, *Eines berühmten Medici gründliches Bedenken und physikalische Anmerkung von dem tödlichen Dampf der Holtzkohlen* (A Famous Physician's Thorough Considerations and Physical Notes on the Fatal Vapors of Charcoal).

Dangers of coal vapors were known of old. Valerius Maximus reports that Hannibal suffocated the inhabitants of Nuceria in baths by vapors and fumes (the first example of "Giftkammern" for mass murder). Coal vapors were used for suicide by Catullus and Seneca. L. Lewin (*Die Kohlenoxydvergiftung*, Berlin, 1920) shows that this knowledge was carried farther by enlightened scientists; Ramazzini also reports occupational poisonings by coal dust. But the people did not recognize the danger, and even physicians looked for other causes of sudden death (noxious emanations from the walls), or attributed it to supernatural forces.

It was Borelli (1608–1697) and G. E. Stahl (1660–1734) who called attention to the danger of coal vapors. Friedrich Hoffmann clearly explained that in a case in which he had to give an opinion, the death of several persons was caused by coal vapors, and he quoted many examples in works published by other authors. He had to defend his stand against another physician, A. F. Erdman, who bluntly declared that the devil had killed the victims for their sins.

Of the publications which dealt with illnesses in all trades and occupations and were more or less based on Ramazzini's work, the first in chronological order is A. F. de Fourcroy's *Essai sur les maladies des artisans traduit du Latin de Ramazzini* (Essay on the Diseases of Artisans, Translated from the Latin of Ramazzini), Paris, 1777. In his additions to Ramazzini he gave some extensive rules for industrial hygiene, especially for gilders. These craftsmen were in great danger from "fire gilding," which was done by laying on gold amalgam (mixture of gold and mercury) and then evaporating the mercury, a process which today is almost completely replaced by gold electroplating. Fourcroy recommends a large workshop with two windows, the room to be used for work only and not for living purposes. The work stove should be opposite the windows, with a large and efficient exhaust pipe. An exhaust hood should cover the whole heated plate with an exhaust tube ending in the chimney or conducted through a window into the open air. The gilder should keep his face turned away from the place where vapors arise. All work, even the scraping off, should be done under the hood. There are also instructions in Fourcroy's book for other artisans.

Ramazzini's work is by far surpassed in its adaptation by Johann Christian Gottlieb Ackermann, *B. Ramazzini's Abhandlung von den Krankheiten der Künstler und Handwerker, neu bearbeitet und vermehrt* (B. Ramazzini's Treatise on the Illnesses of Artisans and Artists reedited and enlarged), Stendal, 1780. Although he omitted almost completely—as he says himself—Ramazzini's numerous argumentations and his poetic figures borrowed from antiquity, Ackermann's book is three times as large as Ramazzini's. Besides many quotations from authors since Ramazzini, he contributes much of his own concerning the control and prevention of diseases. Especially in the chapters on mines and smelting works we find many suggestions, although he is correct in saying that their hygiene is based on adequate installations rather than on medical precepts. He mentions galleries and shafts for ventilation, the blowing in and sucking out of air.

He relates that the Spaniards in America had introduced a kind of work turnover. The Indians work six months in the mines and are then sent back to their tribes, coming back to work after several months. There is also an enumeration of the gases developing in mines, and from his description we are able to recognize carbon oxide, methane, and other gases. If the mines are properly constructed, many accidents could be avoided, he says, by precautions on the part of the miners themselves. Most important is good ventilation. If a miner is missing after a cave-in, his fellow workers attempt to rescue him. Against dust damages, Ackermann recommends—as did Agricola before him—bacon and fatty foods, but work rotation is useful too.

Smelting works should be so situated as to be exposed to winds—not in deep valleys, as is generally the case. Chimneys should be placed so as to produce a strong current of air in the furnaces, so that the vapors are drawn away from the workers.

Other instructions concern other trades; for example, potters should use wet glazes instead of dry ones.

But even in this book directions for adequate nourishment and medicaments to prevent occupational diseases are more numerous than those for industrial hygiene.

THE 19TH CENTURY

Scientific and Economic Evolution. Toward the end of the 18th century a new spiritual, scientific, political, and economic development began, which is not finished today. In the philosophical field a new general view of life was initiated by the French encyclopedists. A gigantic advance in the natural sciences and in technical development followed, from which resulted political revolutions and economic evolutions. It cannot be our task to give more than a sketch of this development; we are interested only in its effects on industrial hygiene.

There are predecessors of capitalism in earlier time. We mentioned above the first steps toward capitalism, in England especially in the wool industry, and in Germany by the Fuggers.

England had already made great industrial strides before

the advent of steam-driven machines. In the first half of the 18th century factories were developed with machines driven by water power. The construction of the spinning jenny was of greatest importance. But it was the use of steam power that made it possible to build factories without regard to local conditions such as water power. Watt's invention of the steam engine in the latter half of the 18th century created the necessary source of power for factories even in cities, and allowed the spread of industrial plants all over the country. In 1810 England had 5,000 steam engines, at a time when on the Continent only a few were to be found.

Lavoisier (1743–1794) made the first chemical analyses. Following him, Wöhler, Liebig, Berthelot and others erected the science of organic chemistry, laying the foundations for the great chemical industry which, through its production of poisonous substances, was to raise many hygienic problems. Later these products penetrated into other branches of industry, even into the smallest workshops and the household, endangering the people.

With this development of industry and commerce the middle class increased their wealth and exerted political influence on legislation as well as government. As owners of factories and all kind of enterprises this class was and still is in sharp contrast to the working class, whose development followed close on its heels. In the quickly growing cities a new class, a mass of workers, was crowded, possessing nothing but their working power, which they had to sell to the class possessing the instruments of work, the factories. For decades this working class lived in greatest misery, in terribly overcrowded houses. Recognizing the harm done to the whole nation by the horribly unhygienic conditions in cities and factories, first physicians, then intelligent and sensible individuals influential in legislature and government, demanded and partly enforced, through legislative measures, an improvement of such conditions. Slowly the economic, intellectual, and moral standards of working men rose. By organizing themselves they learned how to gain better wages and to demand better and more healthful surroundings in houses and in factories. Slowly too their political influence increased, enabling them to influence the legislature. It was a long and precarious process covering many decades.

All these elements, political, economic, and scientific, worked together to build up labor protection and labor protection laws. It is impossible to say which of these cooperating forces—political parties, unions, associations, or scientists—played the most important role in pushing through the various laws or regulations. Other incommensurable forces are mentioned by Collis and Greenwood,[10] in stressing the importance of research work for a certain provision of the Factory Act of 1867. They say: "The code of factory legislation up to the end of the 19th century may not unfairly be said to represent rather an outraged national conscience than to have been based upon scientific inquiry."

So I think it best to begin with the results of all these forces, first showing the development of industrial hygiene legislation, and then studying the forces which worked toward its development.

The beginning and progress of all these economic and legislative changes came at different times and in different ways in the various countries. In France the start was made by a political revolution, which promoted first of all the middle class. Economic changes had begun many decades earlier in England and although changes in hygiene, especially in the hygiene of factories, followed in England only after further decades, they were still earlier than those on the continent, where Germany took the first steps in this

[10] E. L. Collis and Major Greenwood, *The Health of the Industrial Worker* (London, 1921).

direction forty years later. So it is the English legislation that we shall deal with first.

Three publications—one French, one English and one German—show in some respects the tendencies of the new era in legislation and in hygienic-technical installations in factories. These publications may be discussed first.

Patissier, in his *Traité des maladies des artisans d'apres Ramazzini* (Treatise on the Diseases of Artisans, According to Ramazzini), Paris, 1822, has industrial hygiene much more in mind than his predecessors. For the control of industrial damages he recommends:

1. Prohibition of dangerous trades, or, if that is not possible, the exclusive use in them of criminals sentenced to death;
2. Diminution of dangers by substituting machines for hand work;
3. Installation of public baths;
4. If the complete prevention of injury is impossible, the aged and the ill persons should be cared for by mutual insurance.

He also mentions the advice previously given by Gosse (Geneva) that, as protection against noxious gases, workers should wear over mouth and nose a sponge soaked in certain liquids, depending on the vapors against which they are to protect the wearer. He recommends for certain purposes other respirators (which were constructed at the end of the 18th century by Pilatre de Rosier, Aulmayer, and others), but Patissier adds that all such respirators are a nuisance for the worker and should not be used regularly. "We therefore have to look for simple and cheap devices which do not bother the working man and are independent of his volition." As such a device he recommends D'Arcet's "fourneau d'appel." D'Arcet, a tester in the Paris mint, saw three of his seven colleagues die from the effects of nitrous fumes. Seeking a means to draw off the vapors, he introduced the pipe of a special furnace into the chimney of the work shop, which by its heat caused a dilution of the air and so a draft strong enough to carry the fumes out with it. For his invention he won a prize of 3,000 Fr. placed by a manufacturer of gilded bronzes at the disposal of the Academie Royale des Sciences de France. Besides D'Arcet's "fourneau d'appel," Patissier recommended that gilders should wear impermeable gloves and working clothes, and that work clothes and street clothes should be stored in separate lockers. In accordance with the custom in Paris factories of this kind, mirror makers should cover mirrors only one day a week and do other work during the rest of the week. Patissier also speaks very extensively of the sewer workers, who are endangered by "mitte" (ammonia vapors), "plomb" (hydrogen bisulfid), and "gas azote" (lack of oxygen); he recommends various measures for their protection. His chapter on silk workers in Lyon is very interesting; it describes manifold peculiarities, not caused by the work of these tradesmen who, like the "dames des halles" of Paris, are a special class of people in themselves.

In summary, we may say that Patissier's book is the first to offer a modern outlook on industrial hygiene; his acute observations and remarks should be—but are not always even today—recognized by everyone.

From the book by S. C. Turner Thackrah, *The Effect of Arts, Trades and Professions and of Civic States and Habits of Living on Health and Longevity* (2d ed., London, 1832), a few words may be quoted here to show the outlook of the author, worthwhile even today: "Masters, however enlightened and humane, are seldom aware, never fully aware of the injury to health and life which mills occasion. Acquainted far less with physiology, than with political economy, their better feelings will be overcome by the opportunity of increasing profit." He describes how a "diminution of the intervals of work has been a gradual encroachment. . . . Time was thus saved; more work was done; and the manufactured articles consequently could be offered at less price. If one house offered it at a lower rate, all other houses, to compete in the market, were obliged to use similar means. . . . Whatever improvement may be effected without a legislative enactment, restricting the period of labor, this improvement will be but temporary." He regrets deeply the failure of such a bill in the Parliament.

He also writes extensively about bronchitis and the other lung diseases of flax workers.

May I suggest a plan for carrying off the dust? Let channels, about a foot in breadth, be made in the floors, each with one end opening into the room and the other outside of the building. Over the former let a light broad wheel, attached to the machinery, be made to revolve rapidly. A current of air will thus be produced, and this entering the channel, will draw down the greater part of the dust, and carry it out of the building. If the plan succeed in the flax mills, it would avail also for removing the dust of corn and malt mills, indeed of all the manufactures which affect the lungs by mechanical irritation.

A. C. L. Halfort in his *Entstehung, Verlauf und Behandlung der Krankheiten der Künstler und Handwerker* (Origin, Course and Treatment of the Diseases of Artists and Artisans), Berlin, 1845, mentions more extensively than Thackrah, who quotes only a few words of Parkes, the replacement of white lead in house painting by zinc white or white antimonoxid. He reports also on technical installations to control grinder's phthisis. Twenty years ago, he says, a magnetic mouthpiece was recommended, which, if worn during the work, would attract the iron particles. But, he adds, this device did not control the illness (because as we know today, the illness is caused by the quartz particles of the stone, which are also much more numerous than the iron ones). The other device quoted by

Halfort is described first by G. E. Holland [11] in line with the device described by Thackrah. It consists of a wooden funnel about ten or twelve square inches in size, mounted above each grindstone on the side opposite the grinder. From the funnel starts a duct which discharges into a wider duct or canal below the floor and near the wall. This large duct contains a shovel which is connected by a strip with the grinding wheel in such a way that the motion of the wheel moves the shovel quickly, thus causing a strong draft of air in the whole system of ducts which expels the dust through the opening of the duct outside.

[11] G. E. Holland, *Diseases of the Lungs from Mechanical Causes and Inquiries into the Conditions of the Artisans Exposed to the Inhalation of Dust* (London, 1843).

2

Legislation in Europe and the United States

GREAT BRITAIN

Factory Legislation. In all countries the earliest factory legislation pertained to children. Preceding such legislation in England there was one law designed to protect a small group of children whose sufferings were especially terrible. This was the Chimney Sweep Act of 1788. Its purpose was to prevent the employment of children as sweeps, whose lot Percivall Pott had depicted in 1775 in describing the first recognized cancer of occupation.

The first law protecting children in certain textile factories was the Health and Morals of Apprentices Act of 1802, introduced by Peel. It contained regulations for the work of "apprentices" in cotton and woolen factories employing more than three apprentices or twenty other persons. It fixed the maximum hours of work at twelve a day, did not allow night work, and ordered the walls of the rooms in factories to be washed twice a year and the rooms to be ventilated. Besides that, "visitors" charged with the inspection of these factories had the right to "direct the adoption of such sanitary regulations as they might on advice think proper." But visitors were rarely appointed and still more rarely were they active. This law established no restriction as to the age of working children.

It was the Act of 1819 that stipulated nine years as the minimum age of working, limited working hours of young persons, and extended protection to all children, not only apprentices. But it referred only to certain textile factories.

The Act of 1831 widened these regulations to include all textile factories.[1]

It should be stressed that all these laws were passed only after a hard fight. The employers resisted legal action, declaring that it would destroy British industry. They were supported by the economists of that time, who rejected any interference of the state with economics and fought for "laissez faire, laissez aller," that is, for "free enterprise" in no way influenced or hindered by the state. It was a long struggle in which such men as Robert Owen, Lord Ashley, Sadler, Oastler, and Cobbett tried to persuade those of their own economic class that the health and welfare of the people must be placed before money-making. For every new bill the struggle had to be renewed, and even if the principle of the legislative measures was recognized as justifiable, the ensuing law was generally adopted only with serious restrictions which could not be eliminated until many years later.

The consequence of this opposition was that the laws following the Act of 1802 were limited to the work of children (1819, 1825, 1829, 1831). Some of the laws for the protection of children contained regulations which later became helpful for adult workers also. Most important in this respect was the Act of 1833, which enforced the appointment of *factory inspectors.*

The Act of 1844 restricted the work hours of women

---

[1] As I mentioned in my introduction, labor protection as such is treated here only in so far as is necessary to recognize the trend of this development. Readers who are interested in the whole problem may refer to the works of E. L. Collis and Major Greenwood, *The Health of the Industrial Worker* (London, 1921); G. M. Kober, "Historical Review" in his *Industrial Health* (Baltimore, 1924).

Some of the following facts are taken from these two books as well as from Adelaide M. Anderson, "Historical Sketch of the Development of Legislation for Injurious and Dangerous Industries in England" in *Dangerous Trades*, ed. by Thomas Oliver (London, 1902); *Annual Report of the Chief Inspector of Factories and Workshops for the Year 1932, Including a Review of the Years 1833–1932* (London, 1933).

workers of all ages. Until then, the restriction of work hours applied only to young workers. The act granted factory inspectors the power to appoint "certifying surgeons," who were to certify the age of children. It was also the first act making provisions for accident prevention. It prohibited "protected persons" (children and women) from cleaning shaftings and other transmission machinery while in motion and also from working between the fixed and moving parts of any self-acting machine, because self-acting spinning mules were a frequent source of accident. Further it required the secure and permanent fencing of fly wheels, water wheels, wheel races, hoists and teagles which children or young persons were likely to pass, and also of all parts of the "mill gearing" or transmission machinery. Inspectors were authorized to give written notice to an employer that immediate fencing was required for any dangerous machinery. The employer was permitted to refer the matter to arbitration. The act ordered further that any accident must be reported by the employer to the certifying surgeon, who then had to make full investigation of the cause. It ordered also that fines collected for neglect of these rules could be assigned to the injured persons, thus taking the first step toward accident prevention and employers' liability.

Other acts followed in 1845, 1847, 1850, 1853, 1856, and 1862, all of them concerning "protected persons" in protected factories (different kinds of textile factories). It should be added that certain instructions as to cleanliness and sanitation were to be enforced by the sanitary authorities, and in some cases the final decision was left to the courts. The Bakehouse Act of 1863 set up regulations regarding sanitation and cleanliness, to be administered by local sanitary authorities. Bakeries were not included in factory laws until 1878.

All the labor laws issued up to that time protected the

workers of the textile industry exclusively and, for the most part, only certain groups within that industry. It was the Act of 1864 that first included in the factory laws industries other than textile—the manufacture of earthenware and of lucifer matches, for instance. This act brought other progress also; it gave masters the power to establish special rules for dangerous trades, subject to the approval of the Secretary of State. It prohibited the taking of meals in rooms where noxious material was handled. Factories were required to be clean and aired so as to render vapor, dust, and the like, innocuous. By the Sanitary Act of 1866, this regulation was extended to other factories and workshops, with the additional provision that the workrooms should not be overcrowded.

These regulations were taken over into the Factory Act of 1867 and the Workshop Act of the same year, both of which provided that where grinding or polishing produced dust that might be inhaled in unhealthful amounts, the factory inspector could order the installation of adequate devices, driven by mechanical power, to keep the dust away from the worker. The impetus for these regulations was given (according to Collis and Greenwood) by the third and fourth reports of the Medical Officer of the Privy Council, which described the investigations of Greenhow. But—most important—the Act of 1867 finally brought all factories under its scope, and the Workshop Act of the same year made various provisions for workshops, that is, for establishments employing less than fifty persons.

In the year 1867 the workers received the right to vote and used it energetically in 1874. Some minor acts followed. The Act of 1878 consolidated the previous acts. It also authorized the Secretary of State to make provisions concerning cleaning and ventilation, the execution of which he could make the condition for facilitating some protective measures for children and women foreseen in the law. The requirement with regard to the fencing of hoists and teagles was extended to include danger "to any person," not only to protected persons.

The Employers Liability Act of 1880, which extended the liability of the employer beyond the common law, gave employers an incentive to improve safety conditions in their factories and workshops.

The Act of 1883 (An Act to Amend the Law Relating to Certain Factories and Workshops) dealt with white lead factories and bakeries. It stipulated that the operation of a white lead factory was illegal unless certain conditions were fulfilled, among them good ventilation of stacks and stoves, the use of working clothes and respirators, the installation of lavatories and rooms for meals. The Secretary of State had the right to change or increase the conditions. Every employer had to draw up special rules for his own factory, subject to the approval of the Secretary of State.

The Act to Make Further Provisions for the Regulation of Cotton Cloth Factories (1889) set up regulations for temperature and humidity in these factories, the use of wet and dry thermometers, the supply of fresh air, and the right of the factory inspector to require mechanical ventilation for the removal of injurious dust.

The Factory and Workshop Act of 1891 made the requirement as to the fencing of hoists and teagles absolute, irrespective of the passing of persons, and required means of escape in case of fire. Most important, the act authorized the Secretary of State to issue special regulations for machinery and processes in factories and workshops which he considered dangerous. The factory owners concerned were permitted to object and to appeal to an arbitration committee.

That such power was given to the Secretary of State shows clearly that the resistance against every intervention of the government in the management of factories had given way to a better insight and to an understanding that regulatory details are outside the scope of parliamentary discussion. This change was caused, on the one hand, by the growing influence of the working class and by new economic schools which placed first the well-being of the whole people, and, on the other hand, by the understanding of the employers that such regulations and their enforcement standardize working conditions in all factories and hinder a ruthless employer from profiting by substandard installations.

This law laid the foundation for special regulations of certain industries. Because of the numerous technical details involved, such regulations cannot be fixed by laws or elaborated by a Parliament. This must be done by conferences of persons experienced in the particular field. The law came into force on January 1, 1892, and on May 13 the *London Gazette* published the first three regulations concerning dangerous industries: white lead factories, the manufacture of dyes and the extraction of arsenic, and the enameling of iron plates. Five other regulations followed in the same year, seven in 1894.

The Factory and Workshop Act of 1895 brought stricter rules concerning the overcrowding of workrooms, extended to all textile factories the regulations regarding temperature and humidity, including the stipulation of a reasonable temperature in every workroom, and enlarged the right of the Secretary of State to issue regulations which had to be submitted to Parliament. Entirely new was the ruling that required the attending physician to report to the chief inspector of factories every case of poisoning caused by lead, phosphorus, or arsenic and every case of anthrax. Similarly the employer had to report such cases to the certifying surgeon for investigation.

In 1897 came the first Workmen's Compensation Act. This law and the further development of workmen's compensation are so important to industrial hygiene that we shall devote a special chapter to the subject.

In 1898 the first Medical Inspector of Factories, Thomas M. Legge was appointed. Physicians had been engaged in English factory inspection previously but without special medical tasks.

An Act of 1899 ordered retail shops to provide seats for female assistants.

The consolidating Factory and Workshop Act of 1901 gave more detailed rules for safety, cleanliness, space, ventilation, and temperature in factories and workshops, extended the right of the Secretary of State to draw up regulations, and outlined the procedure to be followed. Persons affected were to receive copies of the draft regulations and had the right to register objections. If the Secretary disagreed, he was required to hold a public inquiry at which any person affected could appear. After the inquiry the Secretary of State was permitted to issue the regulation, which was to be laid before Parliament for a given time. The Secretary also was authorized to declare further occupational diseases reportable by the physician.

The first regulations for preventing anthrax infections were published in the year 1899. These concerned skins and hides. Others published in 1900 concerned wool. New regulations followed in 1905 (wool and hairs) and 1908 (East Indian wool).

The Alkali Works Regulations Acts, one of which was passed in 1906, were designed to safeguard the area surrounding factories. The Notice of Accidents Act, 1906, contained new measures covering that field. In 1907 a Factory and Workshop Act brought the laundries under the Factory and Workshop Act of 1901, thus improving the hygienic requirements pertaining to them. In 1908 the use of yellow phosphorus for the manufacture of matches was prohibited by law. An Act of 1911 gave the Secretary of

State the right to set up conditions for cotton clothes factories pursuant to recommendations made by a committee which had inquired into the questions of humidity and ventilation in such factories.

The Police, Factories, etc. (Miscellaneous Provisions) Act of 1916 authorized the Secretary of State to compel the employer by "welfare orders" to provide for the health and comfort of workers: to make available facilities for preparing, heating, or taking meals and to provide protective clothes, first aid arrangements, seats, washing and dressing facilities, and an adequate supply of drinking water.

Later acts restricted still more the work of women and children according to suggestions made by the International Labor Office of the League of Nations.

In 1926 an act gave protection against lead poisoning to persons employed in painting buildings.

In these years, according to the 1932 report of the Chief Inspector of Factories, "we find an increased tendency to prescribe some of the detailed methods (of safeguarding against mechanical and other danger) by Statutory Regulations." Therefore we see new regulations, as those on woodworking machinery (1923) and horizontal milling machines (1928), or amendments to older regulations, such as to celluloid regulations (1921 and 1928), repair of ships in yards (1914 and 1931), the building trade (1926 and 1931). A regulation in 1921 gave prescriptions for the handling of hides and skins to avoid serious anthrax infections. By the end of 1932 some forty-two rules and regulations had been made under the powers conferred on the Secretary of State by the Act of 1901.

The Factories Act of 1937, in addition to consolidating the previous acts, strengthened the provisions for safety and health and the requirements for first aid installations. Section 62 of this act gave the inspector of factories, who formerly was authorized to take samples of substances used only in match factories and in enameling, the right to take samples in every factory where he suspected the possibility of bodily injury.

Other important safety measures include regulations for the construction of machinery. These specify that every set screw, bolt, and so forth, on a revolving shaft shall be sunk and that gearing shall be encased. Any person selling or renting a machine that does not comply with the requirements of the law shall be fined.

If illnesses occurring in a factory were traceable to the nature of a particular job, or if, by reason of changes in any process, the risk of injury arose, the Secretary of State was authorized to require arrangements for the medical supervision (not treatment) of the endangered persons. He also was authorized to prescribe the maximum weights to be lifted, carried, or moved by any class of persons and to require the attending physician to report all industrial diseases, not just those enumerated in earlier laws.

The power of the Secretary of State to make special regulations for safety, health, and welfare was extended, but at the same time he and the chief inspector of factories were authorized to grant exemptions by certificate. In the following years many older rules and regulations were improved, and many new ones and also many "exemptions" were published.

Factory Orders, 1944 edition, published by the Ministry of Labour and National Service, is a book of 388 pages. Under general provisions, it includes five on health (cleanliness, lighting; eight on safety (hoists, chains, and cranes); and three on welfare (first aid). There are also 22 "welfare regulations for particular trades" (cement works, bakeries, laundries, sugar factories, and so on), and 48 "for safety and health in particular trades." Other regulations concern lead paint, the employment of women and young persons in lead processes, notification and investigation of accidents and industrial diseases, and homework.

These rules and regulations show us not only the tendency to make provisions against all the important causes of accidents and industrial diseases but, even more, the tendency to adapt all regulations to the special peculiarities of the work and to make the provisions as clear and definite as possible—not only to say that the rooms should not be overcrowded and the ventilation sufficient, but to explain the meaning of the words "overcrowded" and "sufficient." Such detailed provisions are based on an exact study and knowledge of every type of work and of the possibility of improving it.

A survey of the English factory legislation shows above all its gradual horizontal extension. First children were protected (1802), then young persons (1819), then women (1844). The early laws protected only those working in textile factories; later those working in certain other factories were covered; and in 1867 all factories and workshops were included. These laws protected men only by an occasional hygienic directive. The laws of 1864 and 1867 were the first to contain more extensive hygienic regulations protecting workers of both sexes and all ages. Many of the later acts strengthened and broadened the requirements for hygienic conditions of workrooms in all factories and workshops and made special regulations regarding work dangerous to health. Not only was the sphere of protected persons widened but also the rules were made more specific and consequently more efficient.

The progress of legislation on general factory hygiene may be traced through the following examples:

The Act of 1802 ordered that all rooms were to be limewashed twice a year and duly ventilated. The Act of 1844 protected the workers against humidity. The Act of 1864 required that rooms should not be overcrowded and should be ventilated so as to make dust and gases as innocuous as possible. Several Acts of 1889, 1897, 1898, and 1895 made special regulations for factories with artificially humidified air. The Factory and Workshops Act of 1901, Section 7, directed that: "In every room . . . sufficient means of ventilation shall be provided and sufficient ventilation shall be maintained." It gave the Secretary of State the right to prescribe a standard of ventilation sufficient for any class of factories or workshops. The act authorized the factory inspector to require the use of fans or other mechanical devices wherever damaging dust or gases are generated. It also stipulated that an air space of 250 cubic feet must be provided for each person. The Factory Act of 1937 asked for at least 400 cubic feet for each person, in addition to stipulating that the temperature of every workroom in which workers were seated while performing their tasks must be at least 60°F. (15.4°C.), with a thermometer in every workroom. In this act the regulations concerning cleanliness, overcrowding, temperature, ventilation, and lighting are very extensive, comprising four printed pages.

The Act of 1864 included some definite measures in regard to the "dangerous trades" and tried to initiate further progress. The Act of 1883 was the first to regulate along broad lines the protection of workers in certain dangerous trades. The Secretary of State, to whom certain powers had been given by the Acts of 1878 and 1883, was authorized in 1891 to issue special regulations for dangerous trades. This created for the first time the possibility of making detailed rules for each trade. We can see from the regulations for the grinding trade, for example, how the rules for protection against dust became more and more far-reaching and more and more strict.

In the Factory Act of 1867 the factory inspector was authorized to demand the removal by mechanical means of dust which might be injurious. The same provision is contained in later acts.

Among more detailed regulations in 1909, certain precautions—mechanical exhaust appliances, respirators, and cleansing—were advised in the grinding of metals and racing of grindstones.

In 1925 two regulations were published, one "for the grinding or glazing of metals," the other "for the grinding or glazing or processes incidental to the manufacture of cutlery." These regulated every detail—hoods, ducts, fans, or other devices used in dry grinding—and required adequate exhaust and inlet ventilation and a supply of clean water for wet grinding. The air of the room must be renewed no less than twelve times an hour—in cutlery no less than fifteen times—so as to create a continuous movement of the air. The minimum height of rooms in wet metal grinding factories was established at ten feet, the total window area at no less than one-sixth of the floor area. Other clauses are equally detailed, each regulation comprising six pages.

The progress in industrial hygiene legislation also is indicated by the fact that, in 1883, the hygiene of only two trades—white lead factories and bakeries—was regulated by law. By 1903, 24 trades had been provided with regulations. Today there are 48 regulations for safety and health in dangerous trades and, in addition, several special acts and welfare regulations.

Factory Inspection. All the afore-mentioned laws and regulations would have been useless unless their execution in the factories had been provided for by an efficient system of factory inspection. Factory inspectors were charged with the enforcement of factory laws and had, in addition, the duty of persuading the management and the workers of factories to reduce nuisance. A very important task of the inspectors was to bring about the further development of labor laws by studying the actual working conditions and their effect on health.

In England the first factory "visitors" worked on an honorary basis and were rather inefficient. They were replaced by appointed factory inspectors provided for in the so-called Althorp Act of 1833. At first there were only four inspectors. They were authorized to enter factories at any time and, in the first decade, to make regulations for carrying out the laws, as well as to impose fines in collaboration with a Justice of the Peace. These rights were taken from the inspectors in 1844, because the combination of executive, legislative, and judicial powers seemed an anomaly. The issuance of regulations, first in the restricted sense and later in a wider one, became the duty of the Secretary of State, and fines and punishments were imposed by the Justice of the Peace or by the courts. The Factory and Workshop Act of 1878 created a centralized system of factory inspection with a chief inspector in London. In 1883–1884, there were, in addition to this chief inspector in London, five superintending inspectors, thirty inspectors and ten juniors, distributed throughout the country. In 1893 the first woman inspector was appointed, in 1898 the first medical inspector (Thomas M. Legge), in 1902 an electrical inspector, and in 1903 an inspector for dangerous trades. Each of these individual appointments has since been expanded into a body of several inspectors. In 1910 the authorized staff of inspectors and assistants consisted of 200 persons, in 1939 of 320, in 1944 of 440 persons. In that year 317,040 inspection visits and 63,852 other official visits were made to factories.

A factory inspector is a highly qualified person. He "must be able to 'stand up' to the employer," according to Andrew, and must be a "complete gentleman." No one can become an inspector who is not "a graduate of one of the first class colleges or of a recognized school of engineering or medical science." In 1929, among 29 accepted candidates, 26 had university degrees. A candidate must pass oral and written examinations, attend lectures, and work for some weeks with an older inspector. After two years on probation, the candidate is required to pass an examination on factory law and sanitary sciences. Then he may be appointed. Inspectors must treat as confidential all information concerning installations in factories and may discuss them outside the factory only in official reports and legal prosecution.

Relations between inspectors and management—and workers, as well—have improved during the course of the years. In the first decades, factory inspectors in England as well as in other countries reported much resistance from management, many attempts at deception, and numerous disagreeable incidents. That this occurs even today is evidenced by the fact that in Great Britain before World War II more than 500 firms were penalized yearly for over 1,200 offenses. This represents, nevertheless, a big improvement over the beginning of the century, when there were from 3,000 to 4,000 offenses. On the whole, the factory inspector of today has been aptly characterized by a senior inspector as a friendly adviser, whose advice is backed by the knowledge that he has the power to enforce his recommendations.

As industrial and technical development progressed, processes and machinery became more complicated. This necessitated not only the appointment of specialized factory inspectors, physicians, and electricians but also the cooperation of other scientists. Therefore a Factory and Welfare Advisory Board, an Industrial Health Advisory Committee, and an Industrial Dust Hazard Panel were created in 1943, and an Information Branch headed by E. L. Middleton was set up "for collecting and disseminating information" on all matters of industrial hygiene.

Consultation among the specialists of the individual

branches, employers, employees, and scientists became necessary in drawing up regulations which would provide for the greatest practical safety.[2] Committees and preparatory conferences had to be formed to reach a satisfactory agreement. In 1912 a conference, one of the first in this field, was held for the purpose of obtaining a more uniform standard of safety, particularly in the cotton and wool industries. Since that time many similar conferences have been held. In 1926, a conference of the cotton spinning and weaving trade was called which established, in each division of the trade, a joint permanent committee representing interested parties. These committees are to meet every six months in order to keep the agreements constantly under review.

Similar advisory committees on safety, health, and welfare matters were set up in other industries where Joint Industrial Councils exist. Committees having a more restricted technical basis also were organized, such as one in 1938 to consider practical safeguards for calenders and extruding machines in the rubber trade and another in 1939 to examine the question of fencing heavy power presses. An informal committee investigated the question of avoiding silicosis in steel foundries; it consisted of representatives of employers, trade unions, a technical adviser from Sheffield University, and factory inspectors. As soon as it appeared opportune to give thought to postwar industrial problems, a committee made up of representatives of both sides of the industry and the Department of Factory Inspection was organized in order to obtain advice on the best methods of compliance with the 1937 Factory Act. In 1945 the Joint Advisory Committee of the cotton industry published an interim report on the prevention of mule-spinners' cancer. Other committees of this kind are functioning now. The cooperation between factory inspectors and their de-

[2] H. R. Rogers, "Industrial Advisory Committee" in *Annual Report of the Chief Inspector of Factories for the Year 1944* (London, 1945).

partment or between individual inspectors and such committees has been of assistance to both industry and the department.

As to medical inspectors of factories, the research work performed by them is described extensively in a later chapter. It may be mentioned here that in this field, too, cooperation with consultant physicians of the various branches became necessary. An Advisory Panel on Ophthalmology was created in 1942, and panels on dermatitis and radiology followed.

GERMANY

Factory Legislation. Other countries proceeded with their labor legislation in the same direction—sometimes in similar, sometimes in different ways—in accordance with their industrial development.

On the Continent and especially in Germany the development of industry was slow and was greatly handicapped by the guilds, which were aided by orders and regulations of the government. In Germany the first steam machine was used in 1785 in Mansfeld. In 1790, Goethe made a trip to see the first large steam machine in a mine in Tarnowitz (Upper Silesia). One of the first steam machines was used in Berlin in the famous porcelain works in 1822; in Württemberg the first steam machine was used in 1841.

Siller writes that in the flourishing Saxon textile industry there was no steam machine in 1812. Twenty-nine percent of the spindles were driven by animals, 13 percent by men, the rest by water power.[3] In England, in contrast, the building of water mills had ceased and steam factories were common by the end of the 18th century.[4]

This late development of industry on the Continent

---

[3] Siller, "Gewerbewesen und Gewerbepolizei um 1750 in Preussen so wie die Anfänge des Arbeiterschutzes" (Trade Institutions and Trade Police around the Year 1750 in Prussia and the Beginning of Labor Protection), *Arbeitsschutz*, 1926, p. 1.

[4] C. M. Waters, *An Economic History of England, 1066-1874* (Oxford, 1925).

makes it understandable that in the 8-volume work (1776–1825) of the Austrian scientist Johann Peter Frank, M.D., *System einer vollständigen medizinischen Polizei* (System of a Complete Medical Police), nothing is said about industrial hygiene and workers' health except for some remarks about scaffolds, decaying buildings, quarries and mines, and storage of grain. He mentions an order concerning quarries in Paris, which were to be inspected by an engineer mainly for the protection of the public. He notes that in the mines many installations are necessary and these are not always provided by the operators. Therefore the government must have an exact knowledge of all mines in the country, and "a policy should be introduced, regarding subsurface activities, which should allow mining to be carried on only in accordance with approved principles of construction." If the engineer does not follow them, he should be held liable for any consequences. All that Frank writes about mines fills only two and one-half small pages.

In 1800 Franz Anton Mai, court doctor and professor at Heidelberg, submitted to the Elector Max Joseph of the Palatinate a "bill dealing with the most important chapters in the history of medical police." In the preface to the printed edition Mai summarizes the tasks of intelligent medical police laws under sixteen points; there is, however, no mention of the hygiene of trades or workshops. But he points out that one of the principal tasks of the police physician is "to make known the dangers connected with some trades and to suggest the precautions to be taken and the diet to be followed."

More extensive ideas but without detail are to be found in another book of about the same time, that of E. B. G. Hebenstreit, *Lehrsätze der medizinischen Polizei-Wissenschaft* (Doctrine of the Science of Medical Police), Leipzig, 1791. The author points out that among occupations there is hardly one that does not endanger the workers' health either by virtue of the substances used or worked on, or the strain of the work. "The government should investigate," he writes, "whether or not it is possible to substitute for damaging substances in several trades others less noxious, and thus to protect the workers." [5]

These books show how little value even the best professional men attributed at that time to industrial hygiene in Germany, whose industrial development was more than half a century behind that of England. This also explains why modern labor legislation began so much later in Germany.

There is not only a time lapse between the development of industry and that of labor protection, but also a marked difference between this legislation in England and that on the Continent, especially in Germany. On the Continent the economic theories of Adam Smith and the principles of early capitalism never attained the same degree of influence as in England. The right of the state to intervene in economic life was almost never actually called in question. In England the right to issue laws for labor protection belonged to Parliament, in which workingmen then had no representatives. Parliament reluctantly took the first steps and later, after decades of resistance, gave the right to regulate small and purely technical problems of labor protection to governmental officials under the control of Parliament. In Germany the right of the government and the local administration to interfere in all matters, including economic affairs, had existed since ancient times.

The government used its strength in the 18th century to maintain the guilds and their regulations, going into the smallest details in order to protect the master and the public. On the other hand the government, particularly in

---

[5] A. Fischer, *Geschichte des deutschen Gesundheitwesens* (Berlin, 1933). Other facts also are cited from this book.

Prussia under Frederick the Great (1740–1786), tried to create and to promote industries and manufactures. But these also operated under strictest supervision by the state. In every city there was to be a "factory inspector" who was responsible for the orderly condition of the factories, their technical installations, the methods of work, and the good quality of their products. But nothing is told about the function of this kind of "inspectors" in labor protection.

At the beginning of the 19th century these conditions changed as the ideas of the French Revolution penetrated into Germany in a more or less diluted form, especially in consequence of the Napoleonic wars. In Prussia, men like Freiherr von Stein and Wilhelm von Humboldt promoted the reform. The old organization by guilds was restricted and deprived of much of its power. Through *Gewerbefreiheit* (freedom of trade) the way was opened for the development of industry by the trade laws of November 2, 1810, and September 7, 1811.

But even with these changes and the progress of parliamentary institutions, the influence of the state and of the government had always existed in economic life, and it was only necessary to extend it to the protection of the life and health of workers—here too against the resistance of employers and of the new middle class, which had more or less influence in the Diets of the single states.

The political disunity of Germany, with twenty-one independent states even after the Napoleonic wars and the Congress of Vienna, makes it nearly impossible to write a history of German labor legislation before 1871. It is difficult to give a complete history even of Prussian legislation alone, because not only the government of the state but also the presidents of the provinces and the police of smaller districts and of cities have the power to make regulations. This power they have held since before the time of parliamentarism. We shall point out only a few typical regulations from the period before 1871, first those concerning child labor.

In 1818 the very poor condition of children in the Rhineland, the most industrialized province of Prussia, was brought to the knowledge of the Prussian Minister of Public Worship and Education, von Altenstein. He proposed a law for children's protection, but he could not override the contention that the industry would be weakened in competition with foreign countries. Generalleutenant von Horn reported to the King in 1828 that the youth of the industrial Prussian districts (the Rhineland) were too weak physically to be trained as soldiers. In 1834 the High President of the Rhineland reported on the unsatisfactory schooling of the factory children. A humane factory owner, Schuchard, proposed protection of children in the Rhineland Diet. Finally, in 1839, a regulation was issued by the ministry, forbidding the employment of children under nine years of age and restricting the work hours of young people up to sixteen years to ten hours daily. A royal regulation of 1840 in Bavaria controlling child labor was broadened in 1854. In the kingdom of Saxony the *Gewerbegesetz* (Trade Law) of October 15, 1861, contained a regulation on child labor and also on the health of workers.

Regulations for labor protection were issued as early as the first decades of the 19th century in single states, governmental provinces, and cities. For example, in 1824 the government of the principality of Lippe published a regulation to control dangers in pits for marl, loam, and clay. In 1833 similar directives were issued for the Prussian districts of Königsberg, Köslin, and others. In 1843 and the following years regulations on how to dispose of waste in woolen mills were published in several Prussian districts—Gumbinnen, Potsdam, Frankfurt on the Oder, and others. The district government of Middle Franconia (Bavaria) published rules for mirror factories using mercury, first in 1847, then in 1853 and 1858.

The Prussian Ministry advised the governments of the districts concerning various measures to protect workers [6] —in 1842 on lead poisoning in faience factories, in 1856 on gas works, in 1865 on the installation of factories producing anilin dyes, and in 1871 on the installation of factories for nitroglycerin preparations. The district governments published regulations in accordance with this advice and also independently. The government of Aachen, for example, published a rule concerning the installation of devices for protection against the bursting of grindstones and from the grinder dust in needle factories (1857) and against lead poisoning in lead smelters (1868). The government of Hildesheim published a regulation in 1868 on the installation of factories producing anilin dyes, and in 1875 the government of Düsseldorf issued an order to eliminate the danger caused by the inhalation of grinder dust in the dry grinding of metals.

Regulations of a more general character also were published. A rule concerning the protection of workers in all factories and workshops against dangers to health and life was published by the ministry of Bavaria in 1863, by the government of Düsseldorf in 1874, and by that of Trier in 1875.

An organization of the whole field of legislation for labor and trades was inaugurated by the *Gewerbeordnung* (Act Regulating the Trades) of the North German Confederation, dated June 21, 1869, which was extended to the whole of Germany in 1873. The *Gewerbeordnung* has been changed many times. Part VII, which is especially interesting to us, concerns workers. It was improved in 1878, in 1891 by the *Arbeiterschutzgesetz* (Law Protecting

---

[6] Leymann, "Ein Beitrag zur Geschichte des technischen Gefahrenschutzes in Deutschland, *Zentralblatt für Gewerbehygiene*, XXII (1935), 2.

Workers), in 1897, 1900, 1908, 1918, and 1920. Other laws dealt with special problems, such as the Homework Law of 1914. Section 107 of the *Gewerbeordnung* of 1869 contained approximately the same requirements as were set forth in Section 120 of the later laws, those of 1891 and after: "Employers are required to establish and maintain workrooms, factory installations, machines and tools, and to regulate the work so as to protect the workers against dangers to life and health, as far as possible, in regard to the special trade." Special stress is laid on good lighting, air space, ventilation, and the removal of dust and vapors. The *Bundesrat* (Federal Council)—after 1919 the *Reichsarbeitsminister* (German Minister of Labor)—had the right to issue protective labor rules for the individual trades. If this authority did not issue the necessary regulations, the governments of the individual states had the right to do so for the trades of their state, and the local police could issue rules for their districts. Factory inspectors, introduced into the whole of Germany in 1878, were given the right to order improvements in factories as were based on these rules or on Section 120 of the *Gewerbeordnung*.

The law of 1903 concerning the manufacture of phosphorus matches and similar material is the only law governing a dangerous trade, but many other trades were regulated through rules issued by the *Bundesrat* and the *Reichsarbeitsminister*. The first such regulations, published in 1893, dealt with plants manufacturing lead paints, mirrors, and the production of cigars. They were followed at the end of the 19th and the beginning of the 20th century by a great number of others, including regulations for lead smelters, zinc smelters, quarries, alkali chromates, and Thomas slag. These were improved in the following decades, and others have been added. The revised rules are more comprehensive, more detailed, and more far-reaching than the previous ones.

At the beginning of World War II, 33 such regulations, edited by the *Bundesrat* or the German Minister of Labor, were in force. There were, in addition, many leaflets and instructions about certain dangers and their control. Important also are the numerous regulations published by the governments of the states, especially those of Prussia, Bavaria, and Saxony, and those of district authorities and the local police, the latter often regulating specific industries of certain districts or cities.

A collection of rules issued by the German Minister of the Interior, published in 1915, comprised about 1,400 rules, regulations, and advice concerning labor protection, including work hours but excluding mines. A similar collection that omitted work hours was drawn up by Leymann and published by the German Association for Industrial Hygiene in 1927. It enumerates 863 such rules and regulations. It must be noted, however, that except for the rules edited by the *Bundesrat* (before 1919) and by the German Minister of Labor (after 1919), all of these regulations have legal power only in certain areas, often limited to a small district or a city. As mentioned above, the inferior authorities can issue such a rule only if the higher authorities have not dealt with the same subject. The *Berufsgenossenschaft* (Workmen's Insurance Association) of the trades concerned must be heard before such a regulation is issued. The police have the right to give orders concerning the execution of such laws or regulations in each factory.

Also important for labor protection are paragraphs 16 to 28 of the *Gewerbeordnung,* which orders that permission of the authorities is necessary for the establishment of factories that may be disadvantageous, dangerous, or a nuisance to the neighbors or the public.

We mentioned the *Berufsgenossenschaften*. These are associations of employers of the same trade, prescribed and organized by law for the purpose of compensating workers for accidents and—since 1926—for occupational diseases. They are nonprofit organizations. The *Berufsgenossenschaften* also are obliged to draw up regulations for the control of accidents and diseases within their trade. Commissions composed of employers and employees in equal numbers are formed to discuss and vote on the introduction of such regulations. The rules must be approved by the highest insurance authority, the *Reichsversicherungsamt*. The *Berufsgenossenschaft* is required by law to have all factories and workshops belonging to it supervised by "technical inspectors." In 1936 there were 464 such inspectors, who must cooperate with the governmental factory inspectors. Fines may be imposed for violations of the rules issued by the association. Some of the *Berufsgenossenschaften* give very detailed instructions, such as those of the Association of the Chemical Industry, which fill a 390-page book of small format. These associations play an important role in the control of accidents and occupational diseases.

Factory Inspection. In Prussia the supervision of factories for the purpose of carrying out the regulations of 1839, regarding the work of children, was entrusted to the local police. This supervision proved inefficient. Attempts then were made through local commissions and later (1849) through individual persons, whose principal task was to assist the employers in their business interests but who also had the right to fix working hours in each of the trades. These attempts likewise failed. By the law of May 16, 1853, the government was given the right to appoint factory inspectors to carry through the protection of children. It applied this right to the most industrialized districts: Düsseldorf, Aachen, and Arnsberg. Not until 1855 did these inspectors gain civil service status. They had the powers of the police and the right of access to the factories at any time. Elsewhere in Prussia and in the other German states the police supervised the execution of the labor laws, which were primarily laws protecting children, although there were also health regulations, as we have seen.

After the formation of the North German Confederation, the *Gewerbeordnung* (Trade Law) of 1869 established regulations concerning the duties of the police in labor protection which were valid in the whole territory of the Confederation. It gave every German state the right to appoint factory inspectors, but did not require them to do so. When the German Empire was founded, these regulations were incorporated into the *Reichsgewerbeordnung* (German Trade Law). In Saxonia, government boiler inspectors, who had functioned since 1849, were appointed factory inspectors in 1872 and were charged especially with accident prevention.

In 1878 a change in Section 139 of the Trade Law introduced factory inspection throughout the whole of Germany. Inspectors were to supervise the execution of all laws and regulations affecting labor protection, including accident prevention. Inspectors were given the right by law to enter all factory rooms at any time. They were given the rights of local police but, until 1937, the federal law did not authorize them to issue police orders or to inflict penalties. Compliance with labor regulations was to be achieved by friendly influence over employers. If this failed, penalties were then suggested to the police, or the employer was brought to court. Factory inspectors were required to submit annual reports to the *Bundesrat* (Federal Council) and the *Reichstag* (Parliament). These were published by the Ministry. Gradually the states granted factory inspectors the right to issue police orders (Hamburg in 1898, Prussia in 1909).[7] In 1937 every German factory inspector was given the right to issue police orders for a single factory and also to impose fines.

[7] W. Hatlapa, "Umriss zur hundertjährigen Geschichte der deutschen Arbeitsaufsicht," *Archiv für Gewerbepathologie*, VI (1935), 222.

The number of factory inspectors was gradually increased. In 1891 there were 19 main districts of factory inspection in Prussia; in 1909, 33. By that time the number of inspectors (including assessors) had been increased to 285, by 1912 to 328. In 1928 Prussia had 35 main districts, most of them identical with governmental districts, whose heads were at the same time referees in the corresponding government. They had to supervise the factory inspectors belonging to the main district. In all, there were 164 factory inspection districts, each of these with one leading factory inspector and one or more assistant assessors.[8] In 1940 the number of inspectors and assessors in Prussia was 449; in addition there were eight medical inspectors of factories.

Entrance into factory inspection in Prussia requires the completion of academic study at a technical or other university. The Prussian regulations of 1896 and 1932 demand an additional eighteen months of legal study at a university, especially of the laws important for factory inspection, and eighteen months of practical training in the office of a factory inspector. After this preparatory stage, the candidate has to pass an examination in order to become a factory inspector assessor. An assessor must work with a factory inspector; after several years he is promoted to factory inspector.

Besides this, experienced workers were trained for three or four years in the office of a factory inspector and, after an examination, were appointed to help the inspector in simpler tasks.

The task of a German factory inspector was not only the enforcement of the labor laws in every respect, but also to give expert opinions on the establishment of factories for which a license from the authorities is needed, on boilers, waste waters, etc., and to intervene in labor-management relations and litigations.

[8] *Ibid.*

The development of factory inspection in the other German states was similar to that in Prussia.

WORKMEN'S COMPENSATION

It is not the purpose of this book to deal extensively with workmen's compensation. But because, by making the employer financially interested in prevention, compensation serves to prevent accidents and occupational diseases, important facts of its history may be mentioned.

The Roman law and the European common laws based on it recognized only the liability for damages caused by fault. But centuries ago there existed a feeling among the people that in the most dangerous trade, mining, the employer should bear a certain liability for workers' accidents. So we find as early as 1541, in the mining law of Joachimsthal, an obligation placed on the employer to take care of workers injured in his mine. The feeling that compensation for accidents which arose out of and in the course of employment should not be handled by the common law became stronger as the number of accidents increased, following the ever greater use of machines in industry.

The German *Haftpflichtgesetz* (Liability Law) of 1871 and the British Employers' Liability Act of 1880 increased the responsibility of employers. The first made the employer liable for accidents that were his fault or that of one of his employees. The British act made the employer liable for accidents resulting from defects in plant or machinery due to the master's negligence, or negligently left unremedied by him or caused by the negligence of his subordinates.

The German *Gewerbeunfallversicherungsgesetz* of July 6, 1884 (Accident Insurance Law for Trades and Industry) and the British Workmen's Compensation Act of 1897 introduced compensation for accidents arising out of and in the course of employment regardless of any further circumstances. In the beginning there were a few exceptions, which were dropped by later amendments to the law.

This law differs in many respects in different countries.

The German law created nonprofit organizations of the owners of factories of the same or allied industries, to which each of them must contribute. The compensation is paid as a monthly benefit related to the degree of disability and to the level of the worker's wages. In fatal accidents a pension is paid to the widow and the children. The association also must pay the fees for medical treatment. The duties of the association in accident prevention already have been mentioned. The German law was amended, primarily to extend its efficiency and effectiveness, in 1885, 1887, 1900, 1911, 1923, 1925, and 1934. Most important is the law of 1925, giving compensation for accidents occurring on the way to and from the place of employment.

The British Employers' Liability Act of 1880 was followed by the Workmen's Compensation Act of 1897, which imposed a liability upon the employer to pay compensation on workers' accidents similar to the German law. But it left to the employer the decision as to whether and where he will be insured against the claims of an injured worker. Only in coal mining is there compulsory insurance, which came into force in 1934. For inability to work, weekly payments must be made; for death, a lump sum to the dependents. The British law was also amended often: 1906, 1918, 1923, 1925, 1934. It was entirely rebuilt by the National Insurance (Industrial Injuries Act) of 1946.

The compensation laws became even more important for another branch of industrial hygiene when compensation for occupational diseases was included.

In 1877 Switzerland issued a factory act which in principle treated occupational diseases like accidents, but it was not until 1887 that the necessary rules were issued. England introduced compensation for occupational diseases (the list covered six diseases only) in 1906 after very thorough and extensive investigations. France followed in 1919, Germany in 1926. The International Labor Conference of 1925 proposed an international agreement on compensation for lead and mercury poisoning and anthrax. The Conference of 1935 added further items: silicosis, poisonings by phosphorus and its compounds, arsenic and its compounds, benzol and its homologues, their nitro and amido compounds, halogenated hydrocarbons of the aliphatic series, damage caused by X-rays or radioactive material, cancer of the skin caused by tar, pitch, paraffin, and other substances of this kind. The English list of today comprises 36 points, the German list 23.

For the purpose of compensation an exact diagnosis of every case is necessary. In England and Germany, every person suspected of suffering from an occupational disease and therefore claiming compensation must be seen as soon as possible by an impartial and experienced physician. In Great Britain he must be seen by the "certifying surgeon" of the district, since 1937 called "examining surgeon," and in Germany, according to the order of 1925, by a "fit" physician designated by the governmental insurance office, usually the health officer, but according to the order of 1936, by the medical inspector of factories or a physician designated by him. So the diagnosis is checked, and the factory inspector has a sound basis for demanding improvements in the plant. Hence the value of compensation for the control of occupational diseases is very great.

HEALTH INSURANCE

Without going into details, it is necessary to mention here the importance of health insurance in factory hygiene. This insurance educates the worker to care for his health; it gives him insight into the activity of the physician, the value of the treatment, and also the value of prophylaxis. The benefits enable him to stay at home without starving during his illness, whether or not the illness was occupational in origin. This care during the first stages is the reason that the worst symptoms of chronic poisoning are much rarer in our time.

Health insurance was introduced in Germany by the law of June 15, 1883. It was the first of the great workmen's insurance legislation (sickness, accidents, invalidism, old age) introduced on the initiative of Bismarck, who hoped in this way to win the workers over to his idea of the state. The laws have been improved and extended many times.

Great Britain introduced health insurance and unemployment insurance by the National Insurance Act of 1911 and is now introducing a comprehensive National Health Service.

UNITED STATES

The economic development of the United States differed in time as well as in many other respects from that of Europe.

In 1820, 84 percent of all gainfully occupied persons were employed in agriculture, although large sections of the West did not belong to the United States. The first cotton factory was built in 1790 in Rhode Island. In 1815 about 100,000 persons, most of them women and children, were employed in the textile industry. Craftsmen and mechanics, opposed to the conditions forced upon them by the merchant capitalists, were organized in the Working Men's Party and in trade unions, which had 300,000 members in 1836.[9]

In that year, the Medical Society of the State of New York proposed as a subject for one of its annual prize essays "The Influence of Trades, Professions and Occupa-

[9] Benjanin W. McCready, *The Influence of Trades, Professions and Occupations in the United States on the Production of Diseases*, new ed. (Baltimore, 1943), Introduction by Genevieve Miller.

tions in the United States on the Production of Diseases." Benjamin W. McCready was awarded the prize (1837). In this booklet (newly edited by Genevieve Miller in 1943),[10] he describes how difficult it is to obtain material on the subject in this country, because workmen pay "so little attention to the causes which affect their health, and their views are so often warped by prejudice or interest." He drew freely from the books of various unions and societies of mechanics, and from two American biographical dictionaries for professional data. Because of the scarcity of material he leaned heavily on the book by Thackrah, published a few years earlier in England. It is interesting to note that he reports—in sharp contrast to reports from Europe—the strict morals of women workers and their piety. Instead of strengthening their health by adequate exercise, the seamstresses hurried to prayer meeting or lectures.

It seems that the interest of the Medical Society of New York in workers' diseases had been aroused partly by Thackrah's book and partly by the formation of the Working Men's Party. For many decades after the publication of McCready's booklet we find no scientific book in this field. There are, however, popular writings. J. Brown, M.D., published a small booklet, *Health, Five Lay Sermons to Working People* (New York, 1865), in which he states that diseases are caused by heredity as factors and also by hard work. The lead colic of painters is mentioned. D. F. Lincoln, M.D., published *School and Industrial Hygiene* (Philadelphia, 1880), devoting 40 of its 105 pages to the chief problems of industrial hygiene: grinder's asthma, rheumatism, caisson work, miners, overstraining, accidents, statistics. His book is by no means free from errors.

A more important scientific work is contained in the *Cyclopaedia of the Practice of Medicine*, edited by Dr. H.

[10] *Ibid.*

von Ziemssen, professor in the University of Munich, "under the immediate editorship of Albert H. Buck, M.D., New York" and published by W. Wood in 1879. Volume XIX, a supplementary volume *On Hygiene and Public Health,* contains an interesting chapter, "Hygiene and Occupation" (78 pages), written by Roger S. Tracy, sanitary inspector of the New York Board of Health. Tracy writes that the environment of workers in factories had improved but new dangers had developed, specifically caused by chemical products. In Massachusetts, children under ten years of age were forbidden to work; in Rhode Island, children under twelve years; Pennsylvania, thirteen years. He urges that work be prohibited during puberty. In addition to many quotations taken from German books by Hirt and Zenker, there are observations of his own. He discusses the damage caused by insufficient time for meals, by workers "bolting their meals and gulping down hot drinks as if for a wager. No more saddening and ludicrous sight can be seen than a lunch counter in the business part of New York from 12 till 2 P.M."

There is also a chapter of 22 pages on the "Hygiene of Coal Mines," written by H. C. Sheafer. He reports on ventilation by the use of a "brattice," an airtight partition carried along one side of the gangway. Ventilation was also achieved by means of natural forces or of furnaces at the foot of the upcast shaft. The first method is not effective in deeper shafts; the latter may cause explosions. Consequently in recent times exhaust fans have been used. According to the report of the Pennsylvania Anthracite Mine Inspectors, from 1871 to 1877 the average number of miners killed annually by accidents was 240, or 0.35 percent. Of 185 accidents, ten were caused by carburated hydrogen gas. Another chapter in the same work, written by R. W. Raymond (twelve pages), deals with metal mines.

Interest in occupational diseases and dangerous trades

arose in the United States at the beginning of the twentieth century. Carroll D. Wright, Commissioner of the United States Bureau of Labor, charged Dr. C. F. W. Doehring with responsibility for investigating the manufacture of white lead, paints, fertilizers, and other substances. The results were published in 1903.

In 1908, as chairman of the Committee on Social Betterment of the President's Homes Commission, Dr. G. M. Kober submitted a report on Industrial and Personal Hygiene. In 1908–1909 the Bureau of Labor published Dr. Frederick L. Hoffman's monograph, *The Mortality from Consumption in Dusty Trades*. John Andrews, the promoter of the American branch of the International Association for Labor Protection, was charged by Commissioner Wright with the task of carrying on an investigation concerning phosphorus necrosis in the center of the match industry. He found 150 cases, among them serious ones. This was followed by the Esch Law (1912) which imposed such a high tax on white phosphorus matches that their production became unprofitable. In 1910 Dr. Alice Hamilton was asked by the Commissioner of Labor, Charles O'Neill, to investigate the lead industry throughout the country, which she did in the following years.

In 1911 the July *Bulletin* of the United States Bureau of Labor published a comprehensive treatise by Thomas Oliver (England) on industrial lead poisoning, which was followed by an article by Alice Hamilton on "White Lead Industry in the U.S., with an Appendix on the Lead Oxide Industry" and one by J. B. Andrews on "Deaths from Industrial Lead Poisoning (Actually Reported) in New York State 1909 and 1910."

In the second half of the first decade of this century several states also began to initiate investigations into factory conditions, especially in dangerous trades.

In 1906 the Massachusetts Board of Health issued a report on the conditions affecting health and safety in factories. The report was supplemented in 1907, partly as a result of investigations regarding dusty trades made by W. C. Hanson.

In 1908 and following years, the New York State Factory Investigating Commission made extensive inquiries, which we shall discuss later.

The state of Illinois appointed an Occupational Diseases Committee in 1908 which included five physicians, some of whom later became authorities in this field, particularly Alice Hamilton and Emery Hayhurst. The former studied lead, the latter copper, while Peter Bassoe examined caisson sickness, of which there were 161 cases in the construction of tunnels in Chicago.

Formation of these committees was indicative of the increased interest in industrial hygiene, and also encouraged the enactment of legislation. In addition to these official organizations, private organizations were founded with similar aims.

In 1906 an American section of the International Association for Labor Protection was created which established a permanent secretary's office in New York. John Andrews served as secretary. In 1910 publication of a regular bulletin was begun.

In 1910 the First National Conference on Occupational Diseases was held in Chicago. In 1911 an American Museum of Safety was founded in New York by W. H. Tolman. In the same year the National Safety Council was set up in Chicago. At the International Congress of Hygiene and Demography held in Washington, D.C., in 1912, a large section was devoted to industrial hygiene.

During these years, conditions in the garment industry of New York were investigated, following a strike in this industry. G. M. Price published the results, which will be dealt with in a later chapter.

Factory Legislation and Governmental Activities. Although legislation of the separate states is of more importance in the practice of factory hygiene and preceded federal legislation in most cases, nevertheless the research, investigations, and publications of the federal bureaus and departments have great theoretical value and exert a farreaching practical influence on the legislation of the states.

Federal Legislation and Activities. In labor questions, the acts of the federal government and the Congress apply only to its own employees, to persons engaged in work financed by federal funds, and to workers engaged in "interstate commerce or in the production of goods for interstate commerce." Most of these acts refer to workers in interstate transportation. But a few go farther, as, for example, the Act of 1890–1891 dealing with mines, the Esch Law of 1912, and—most important—the Fair Labor Standards Act of 1938.

Between 1885 and 1888, several acts were passed which regulated the working hours and holidays of governmental employees. An act of August 19, 1890, provided that apprentices for sea duty must be not less than twelve years of age, must be in good health and of sufficient strength. An act of 1890–1891, concerning hygiene and safety in mines, prohibited the underground employment of children under the age of twelve. In 1893, safety devices were required on railroads. In 1897, an inspection of boats propelled by gas was ordered. From 1899 to 1912, several acts were passed pertaining to railroads and vessels. The Esch Law, which prohibited the use of white phosphorus in the match industry, was passed in 1912. Statutes and regulations for safety appliances on and inspections of railroads followed later. These were intended for the protection of the public as well as of the employees. An act of 1935 regulates the hours of service of motor carriers.

Most important is the Fair Labor Standards Act of June

LEGISLATION IN EUROPE   57

25, 1938, which applies to workers in interstate commerce or in the production of goods for interstate commerce. The Act set minimum wages at thirty cents an hour up to October, 1945, and at forty cents an hour thereafter. Work in excess of forty hours a week must be paid at the rate of time and a half. No goods for interstate commerce may be produced by children under sixteen years of age, or under eighteen in certain dangerous occupations. The law is administered by the Department of Labor.

The Act of June 23, 1938, for first pilots in Civil Aeronautics, set daily labor hours at eight hours in twenty-four, or thirty hours in seven days, and established a minimum salary. There are, in addition, laws relating to social security, industrial relations, unemployment insurance, old age and survivors' insurance, public assistance to the needy, and health and welfare services, including maternal services and others. The Employers' Liability Act covers employers of railroads engaged in interstate or foreign traffic, and applies also to seamen. The Seamen Laws, too, contain hygienic provisions. During World War II, the Walsh-Healey Act empowered the Secretary of Labor to draft regulations for work under governmental contracts, and the National Committee for Conservation of Manpower created a safety organization throughout the forty-eight states.

The report of the Committee on Safety and Health and Workmen's Compensation of the 12th National Conference on Labor Legislation (December, 1945) expresses the regret that, "Federal emergency safety agencies have been withdrawn with the termination of hostilities. The major burden of preventing job accidents and disease now rests with the States. Many of them assume this task with inadequate tools. . . . Few labor departments have been able to obtain sufficient State appropriations to recruit qualified staffs to do the job." The committee, therefore, urges federal aid to states in the enforcement of state labor laws. It also urges that federal funds be provided for the safety training of workers, and asks for the cooperation of trade unions. The tenor of the conference is an urgent desire for greater federal influence and activity.

Federal laws, especially the last mentioned, influenced the activity of state legislators and the development of state laws. Much greater is the influence of research studies, investigations, advice, and education carried on by the various federal departments. Although federal legislation is much restricted by the Constitution, and legislation in some states lags, no country in the last decade has done so much and invested so much money in industrial hygienic research as the United States. This is due in some measure to the Divisions of Industrial Hygiene maintained by a few states, but far more largely to federal institutions.

Efforts to create a federal Office of Labor began in 1871, but it was not until 1884 that an "Act establishing a Bureau of Labor in the Department of the Interior" was signed. The purpose of the bureau was to collect information on labor and hours of work and also on "the means of promoting the material, social, intellectual and moral prosperity" of workers. Dr. Carroll D. Wright was appointed the first Commissioner of Labor. In 1888 the bureau was made independent of any other department, but in 1903 it became attached to the Department of Commerce and Labor. Since 1913 it has functioned separately as the Department of Labor. From 1886 to 1921 it published 25 annual reports, 50 miscellaneous reports, 13 special publications, and 291 bulletins. The 1945 list of governmental publications on labor is a booklet of 91 pages. Today the Department of Labor has a Division of Labor Standards, a Wage and Hours and Public Contracts Division, a Conciliation Service, a Bureau of Labor Statistics, a Children's Bureau, and a Women's Bureau. Through cooperation with the indi-

LEGISLATION IN EUROPE

vidual states, the department and its divisions attempt to bring about uniformity of laws and regulations.

The United States Public Health Service was created by the act of August 14, 1902, the Division of Industrial Hygiene in 1915. The Public Health Service cooperates with industry, the universities, and many other organizations. In addition to carrying on practical and research work, including numerous research studies in industrial hygiene, it also provides for the instruction of engineers and physicians. Since 1915, 76 major publications on industrial hygiene have been published in the Public Health *Bulletin* (see Chapter 8), and many smaller ones in the Public Health *Reports* and other journals. In many states, "Industrial Health Units" were installed with the cooperation of the Public Health Service; these publications and installations will be discussed later.

The National Bureau of Standards, in existence since 1900 as a part of the United States Department of Commerce, has published information on X-rays and radium protection, rules for the installation of electrical equipment, and similar safety advice.

The United States Bureau of Mines was created by the law of May 16, 1910, and its statutes improved in 1913. Its earliest work on hygienic questions was done in cooperation with the Public Health Service, whose physician, A. J. Lanza, together with Edwin Higgins, was the first to study silicosis in the United States (in the mines of the Joplin region, Missouri). In 1920 the post of chief physician was created and R. R. Sayers was appointed to the position. In 1926 a Health Division was founded. Its activities are discussed further in Chapters 8 and 10.

In the field of labor protection, the following federal authorities are active today: the Department of Labor with several divisions, the Department of the Interior with its Bureau of Mines, the United States Security Agency with its Public Health Service and Office of Industrial Hygiene and Sanitation.

State Legislation and Activities. The economic conditions and industrial development in the different states vary according to the history of their colonization by the white man, as well as their size, the number of inhabitants, the quality of the soil, the climate, and occupational possibilities. A few statistics may suffice to illustrate this. In 1940 the state of New York had 12,879,646 inhabitants and Pennsylvania 9,900,130, while Rhode Island had 713,346, Delaware 266,505, Nevada 110,247. The population per square mile in Rhode Island is 674.2, in Massachusetts 545.9, in New York 281.2, in Ohio 168.0, in Wisconsin 57.3, in South Dakota 8.4, in Nevada 1.0. The urban population of Rhode Island comprises 91.9 percent of the entire population, in New York 81.6 percent, in Arkansas 22.2 percent, in North Dakota 20.6 percent. In view of such variety in size and density of population, as well as in industries, one cannot expect uniformity of legislation and institutions, and there are indeed great differences.

Whereas in England and on the Continent there were officials or governmental institutions which for centuries had cared for economic or trade questions and which were able, with the development of industry, to take over the new tasks, in the United States such institutions had to be created. To investigate all the problems of labor, of the health of workers and of labor-management relations, and to provide legislative bodies with the necessary material, new institutions had to be organized.

The first of them, the Bureau of Labor Statistics, was created in 1869 in Massachusetts in consequence of agitation by a workers' organization, the Order of the Knights of St. Crispin. Pennsylvania followed in 1872, Kentucky and Texas in 1876, Ohio in 1877, New Jersey in 1878, Illinois and Indiana in 1879, New York, Michigan, and others in 1883. By 1899, there were 30 such bureaus; by 1910, 34. Today every state has a bureau or department or board of this kind. In addition to their fact-finding duties, they also handle all matters relating to industry and workers and have the task of enforcing and administering all labor laws. These institutions differ widely in size and organization. The Labor Department of the state of New York has, besides the Division of Industrial and Safety Standards, one of Industrial Relations with a Bureau of Women in Industries, and a Division of Research and Statistics, all these with various sections and units. The Department of Labor and Industry of Massachusetts has, among others, a Division of Industrial Safety, one of Statistics, one of Standards, and one of Occupational Hygiene.

On the other hand, to mention the smallest organizations, Arizona has an Industrial Commission of three members responsible for the enforcement of all labor laws. But there are no laws protecting workers' health, except regulations concerning work hours, sanitation, and ventilation in laundries. In Kansas, the Labor Department has one section for factory and mill inspection and another for mine inspection. There are, however, no health laws which require enforcement.

In many states, bodies of this kind are also authorized to establish rules and regulations for the safety and health of workers. In nineteen states they have no such power; all rules and regulations must be passed by the legislature. As was pointed out previously, detailed specifications for the health of factory workers, especially in dangerous trades, do not lend themselves readily to legislative discussion. So it is quite understandable that such laws do not exist in some states.

Of all divisions and boards of the labor departments and similar organizations, most important for our field are those whose task is the protection of the workers' health.

Two states, New York and Massachusetts, have in their Labor Department a Division of Industrial Hygiene. Illinois has an Industrial Hygiene Section, headed by a physician, within the Division of Factory Inspection.

Farthest advanced in this field is New York State. I shall trace the development in this state in some detail because there is, to my knowledge, no similar organization in any European state.

A Bureau of Labor Statistics was created in New York in 1883, and the inspection of factories was begun in 1886. In 1906 Chief Inspector Sherman came to the conclusion that expert scientific knowledge was "indispensable to the proper experience of the function of the Bureau of Factory Inspection." In the year 1907, C. T. Graham Rogers was appointed Medical Inspector of Factories. In 1908 J. Williams recommended the appointment of a competent engineer "to pass upon plans for ventilation, proposed as a result of orders issued by the Department." In 1910 a chemist was appointed and a laboratory placed at the disposal of the department. A mechanical engineer was appointed in 1912 to give instructions on the safeguarding of machines as required by law. In 1911 a big factory fire resulted in the creation of a Factory Investigating Commission to investigate the conditions under which manufacturing is carried on, including matters affecting the health and safety of operatives. The commission appointed as head of the work of inspection and sanitation Dr. George M. Price, who organized a corps of inspectors for this purpose. Many public hearings also were conducted.

The preliminary report issued in 1911 contained as Appendix VI a report of 197 pages on occupational diseases, lead and arsenic poisoning in the city of New York, by E. E. Pratt. A second report followed in 1913, a third in 1914, and a fourth in 1915, twelve volumes in all, including nine volumes of protocols of hearings. The commission reported the "lack of a sufficient number of inspectors, especially inspectors having scientific and technical training" and recommended the creation of a Division of Industrial Hygiene for the control of occupational diseases and industrial accidents and an Industrial Board for the promulgation of rules for the protection of the health of the workers. In 1913 the Division was created, a laboratory was equipped, and inspectors were assigned. In 1914 the personnel consisted of four physicians, one chemical and one mechanical engineer, a civil engineer, and an expert in fire prevention; but in 1916 the staff was reduced to two physicians, one chemical engineer, and two inspectors.

During the following years, in addition to the routine work, publications were issued which dealt with exhaust systems and arsenic poisoning. The position of the Division in the Labor Department varied from time to time, but it worked always in the same direction, doing practical work in improving the installations in factories, especially exhaust equipment, and carrying on educational and research work. In 1924 the publication of an *Industrial Hygiene Bulletin* was begun. In 1931 it became a part of the *Industrial Bulletin* of the Labor Department, but now is published separately again as the *Monthly Review, Division of Industrial Hygiene*.

A degree of cooperation was initiated between the Division and the Workmen's Compensation Board in 1925. Dust and lead, carbon monoxide, luminous paints, caisson work, and other problems were studied and the results published. Education by lectures, radio, and an exhibit were continued.

In 1936 Leonard Greenburg became Executive Director of the Division and intensive studies were conducted into silicosis (foundries) and chlorinated naphthalins. Older codes concerning rock drilling were improved and an X-ray truck was made available for examination of dust workers.

The personnel has been increased to 34 persons, the laboratory enlarged and better equipped, and the activity extended in every direction.

Today the Division of Industrial Hygiene and Safety Standards comprises four units which work very closely together: the medical unit (five physicians), the chemical (six chemists), the engineering (twenty engineers), and the code unit, transferred to the Division in 1946. The work of the Division falls into two main categories: a technical consulting service to the rest of the department, and its own industrial health program. In 1946 the Division of Industrial Safety Service (factory inspection) referred 2,261 cases to the Division of Industrial Hygiene, which analyzed for it 461 field samples, including 335 chemical air analyses. The Board of Standards and Appeals needed the advice of experts to determine whether a proposed order for crushing plants was justified. The Workmen's Compensation Board requested 111 investigations.

The industrial health program includes the detection, control, and prevention of occupational diseases. Among research studies conducted last year were investigations of caisson illness, exposure to radium, silicosis in the ferrosilicium manufacture, and methylchloride. The engineering unit cooperates with private industry by testing and approving building and ventilation plans, some of which are submitted to it in the blueprint stage. The Division also introduced plant medical services, studied adequate nutrition in factory canteens, and continued its educational work.

The extent of the activity of the Division is indicated by the following statistics: in 1946 the medical unit took 19,758 micro chest films, made 502 plant visits to study prevention, and analyzed 1,165 compensation files. The chemical unit made 2,314 chemical determinations, analyzing 341 chemical substances and 963 air samples, and visited 383 plants.

The engineering unit visited 2,230 plants, acted upon 2,587 exhaust plans submitted, examined 2,921 building plans, and held 1,913 conferences with architects, engineers, and others. The code unit worked on 15 codes.

In addition, there was the publication of the *Monthly Review* and of articles in other journals, amounting altogether to 30 articles in 1945. Up to the end of 1945, 133 articles written by staff members had been published in the *Review* and the *Bulletin* and in other publications, including special bulletins on almost every problem. Of particular importance were articles on silicosis (L. Greenburg, Theodore Hatch, A. R. Smith), on carbon monoxide (M. R. Mayers, William J. Burke), on lead (M. R. Mayers), and on the fur industry (H. Heiman, M. McMahon).

There is also in the Labor Department a Division of Industrial Safety "which is responsible for the enforcement of laws and rules" and makes factory inspections. Created in 1886, its duties have been increased frequently. Since 1921 there have been several different divisions of factory inspection. In addition to 51 higher employees—a director of inspection, chief factory inspector, supervisors, and their assistants—there were in 1946, 147 factory inspectors, 51 inspectors for construction work, 10 for boilers, and 4 for mines and tunnels, making a total of 51 supervisors and 312 inspectors. The salaries of the higher employees range up to $8,000, those of the inspectors from $2,200 to $3,600 yearly. Nearly all are civil service employees.

The Division of Occupational Hygiene of the Labor Department of Massachusetts is also very active; it has published a number of valuable articles, including those of M. Bowditch and H. H. Elkins. In Illinois, Dr. Strauss of the Industrial Hygiene Section of the Labor Department's Division of Factory Inspection has written several articles for the *Illinois Labor Bulletin*.

In New York and Massachusetts, the Division of Industrial Hygiene is incorporated in the Labor Department, which alone has the authority to enter factories and enforce labor laws and regulations. The divisions work closely with other divisions of the department, especially that of factory inspection.

Recently, 55 industrial health units have been installed with the aid of the U.S. Public Health Service and with the financial support of the federal government.[11] Of these units, 43 are in State Health Departments, one is in the District of Columbia, 4 are in territories, and 8 in local health departments. They employ 308 professional persons, about 60 percent of them engineers and chemists, and 21 percent (75 persons), physicians and nurses. Only about 27 percent of the units offer medical consultant service. None of these units have the right of the State Labor Department to inspect factories and enforce rules or recommendations. Their only activity is as consultants for the industry. The 42 units which reported for the year 1944–1945 to the U.S. Public Health Service listed 23,725 instances of services to 16,205 plants. The overwhelming majority of these services constituted determinations of air contaminants, and laboratory examinations of materials and other substances.

State Laws, Codes, and Regulations.[12] The earliest state laws for the protection of workers concerned children and women. These laws were followed rather late by studies in and laws concerning factory hygiene and the prevention of occupational diseases. States along the eastern seaboard led the way. In 1827 a child-protection bill was drawn up

[11] V. M. Trasko, "In Industry's Service," *Occupational Medicine, III* (1947), 392–405.
[12] Sources consulted include: G. M. Kober and E. R. Hayhurst, *Industrial Health* (Baltimore, 1924), particularly the chapter, "Historical Review"; G. M. Kober, "History of Industrial Hygiene" in *A Half Century of Public Health*, ed. by Mazych P. Ravenal (New York, 1921); J. B. Andrews, *Administrative Labor Legislation* (New York, 1936); and J. B. Andrews, *Labor Law in Action* (New York, 1938).

in Pennyslvania but was rejected by the senate, and no such bill was passed there until 1848. The first child labor law went into effect in Massachusetts in 1842. Most of the other states followed at long intervals—New York and Illinois in 1886. As late as 1890, only 34 of the 54 states and territories set a minimum age for children, and even in 1946 Wyoming had no such law.

By 1890, 21 states had statutes relating to accident prevention and the health of employees, in addition to statutes on the employment of women and children. Among them were laws concerning ventilation in Massachusetts, New Jersey, Ohio, New York, Connecticut, Pennsylvania, Rhode Island, and Missouri; on heating and lighting in New Jersey and Ohio only; on lime washing and painting of walls in New York, New Jersey, and Michigan. Nineteen states required that seats be provided for women, among them—strangely enough—Alabama, Colorado, Kentucky, and Mississippi, which had no child labor laws. Toward accident prevention, eight states had laws which prohibited women and children from oiling machines that were in motion, and ten states had laws on the guarding of elevator and hoistway openings.

By 1920, 36 states had laws requiring guards for dangerous machinery, and 35 states had laws concerned with health and sanitation provisions, some of which require the removal of injurious dust and fumes by mechanical devices and exhaust ventilation. Since that time the contents of labor laws and the number of laws, codes, and regulations have increased continuously. Some states have numerous and well elaborated rules, also called regulations or codes. Only a few examples can be cited here.

There is no sharp dividing line between the content of labor laws and that of rules. Some labor laws include an article on the general duties of employers to protect the health and safety of employees. In the New York Labor

Law, for example, §200 reads: "All places to which this chapter applies shall be so constructed, equipped, arranged, operated and conducted as to provide reasonable and adequate protection to the lives, health and safety of all persons employed therein. The board shall make rules to carry into effect the provisions of this section." In addition to these "General Duty" articles, the law contains detailed and specific directions, such as in §202, which provides for the "protection of persons engaged at window cleaning" (in public buildings only). Paragraph 203 requires washrooms and water closets for elevator operators; other paragraphs prohibit the eating of meals in workrooms where poisonous substances, fumes, and dust exist; protect workers in building construction, demolition, and repair work; and list specifications for safeguarding machinery, for illumination, fire doors, fire escapes, cleanliness, closets, and dressing rooms. Article 15 concerns mines, tunnels, work in compressed air (with stated decompression periods), and the use of blasting explosives.

Similar general duties are set forth in the Alabama General Laws (Section 17) and in the Labor Laws of Georgia (Section 10).

The Labor Laws of Nevada contain no such paragraph on general duties, but included in the laws are many detailed directions. The laws also establish the right of the Industrial Commission to fix reasonable standards and prescribe orders for the installation and use of safety devices.

Peculiarities exist in the laws of certain states. Article XVI, Section 4, of the Constitution of Montana provides for an eight hour day in all industries and occupations except farming and stock raising. The legislative assembly "shall have no authority to increase the number of hours" but may reduce them. Paragraph 3090 of the Labor Law stipulates "equal pay for women for equal service."

The Labor Law of South Carolina (§1272) declares it unlawful in the manufacture of cotton textiles "to allow or permit operatives, help and labor of different races to labor and work together within the same room," to use the same doors at the same time, or to use the same pay windows or the same stairway at the same time.

The Labor Law of the Commonwealth of Virginia includes a few health orders, among them a rather extensive one for foundries and another for grinding and polishing wheels. Section 1821, on the "protection of employees of peanut cleaning establishments and cotton factories," asks only that "a suitable sponge shield" be provided to protect the workers from inhaling dust. These sponges must be furnished by the employer at actual cost and must be paid for by the employee. Seats for females are required in establishments other than fruit and vegetable canning factories—one seat for every three females—and their use "shall be allowed at such time and to such extent as may be necessary for the preservation of their health."

North Carolina's labor laws contain more extensive orders. They provide for a Bureau of Labor for the Deaf, seats for women, and a "medical chest." From all these examples we can see that some labor laws include details which one would expect to find, rather, in regulations. But none of these laws contains, and none could contain, all the details necessary for an adequate regulation of all the different industries in an industrialized state. The most advanced states, therefore, supplement the laws with codes. New York, for example, has, in addition to its detailed laws, 37 codes. These codes deal in part with special problems (automatic sprinklers, fire alarm systems, and lighting) and in part with special types of work (foundries, laundries, the needle trades, and canneries). The rules are established by the Board of Standards and Appeals, which consists of at least three members, appointed by the governor with the consent of the senate. One of the members must be a licensed engineer with at least ten years of practical experience. Any person who finds that a code involves him in practical difficulties or unnecessary hardship may petition the Board for a variation; these codes are easily changed.

Rules issued by the Industrial Commission of the State of Illinois cover a large field. Separate sections are devoted to safety measures for the prevention of personal injuries (guarding of transmissions, prime movers, and machinery), to rules relating to the removal of dust from grinding and similar operations, and to others relating to the construction of underground tunnels. Other states, however—Arizona and Kansas, for example—have as yet no rules for the health of workers.

It would take too much space to discuss in detail the codes and regulations of every state, which, though they deal with the same subject, sometimes differ in a few points. In general, these regulations may be compared with those of European countries (some apparently had their source in the English regulations), although such provisions as a periodical medical examination of endangered workers seem to play a smaller role in the United States than abroad. On the other hand, much importance is ascribed in this country to the chemical control of the air and its content of noxious gases. Some states establish "allowable limits," that is, the maximum contents of noxious gases allowable in the air.

And yet, even at the 26th convention of governmental labor officials, held in New York City in 1940, J. M. Falacz of the Illinois Department of Labor could say: "We know further from a study that many States in the Union do not possess any enforceable rules or laws pertaining to health and safety of workers."

These great differences—some states with numerous and

well-elaborated laws and rules, others without any regulations concerning the health of workers—are understandable when we consider the vast economic differences between the states.

Factory Inspection. The enforcement of the child labor law in Massachusetts was entrusted to a single deputy detailed by the Police Department in 1866, but it was not until 1877 that the right of access to factories was given to this inspector. A system of factory inspection was adopted by New Jersey and Wisconsin in 1883, Ohio in 1884, New York in 1886, Connecticut and Maine in 1887, Pennsylvania in 1889, Illinois and Michigan in 1893. By 1894, fourteen states had laws providing for the inspection of factories; by 1920, fifty states and territories. But neither the training nor the position of most factory inspectors was satisfactory. The statement of a leading factory inspector in 1940 illustrates the situation: "No factory inspection division expects factory inspectors to be chemical engineers, chemists, ventilating engineers or physicians." In European countries, however, this is what factory inspectors are required to be, and United States federal inspectors of government work must be engineers with some years of practical experience.

The lifelong economic security which the European factory inspectors have by virtue of being government employees gives them the necessary independence from private interests. But in the United States, inspectors in only some of the states—and not all inspectors even in these states—have civil service status and thus enjoy a certain independence. In 1935, according to Andrews, factory inspectors in only fifteen states were civil service employees. Andrews also noted that salaries were low—$3,000 at most. Now they are up to $3,600.

Several states require boiler inspectors and medical inspectors to have preparatory training. In 1940, an advisory committee to the United States Secretary of Labor composed of high Labor Department officials and representatives of labor organizations prepared a bulletin entitled *Qualifications for General Labor Law Inspectors* (Bulletin No. 38), in which they set forth detailed proposals. The foreword reads in part: "Salaries and opportunities for promotion are seldom such as to retain effectively the services of competent, trained inspectors and in many cases even such inspectors are removed involuntarily from the public service through the vagaries of personal or partisan replacement." It is revealing that among the proposed duties of these inspectors is the enforcement of laws or regulations relating to hours of work and child labor, but nothing is said about hygienic and safety conditions. The applicant is required to have had at least five years of full-time paid employment in business, industry, or professional fields. College or university studies may be credited toward this experience to a maximum of three years, graduate studies in economics up to one year. Basic qualities desired include integrity and tact. The written examination must indicate the applicant's ability to prepare written reports and to make such mathematical computations as are necessary in the examination of payrolls. Training should include a probationary period of at least six months, not less than two weeks of specific instruction, and four weeks of work with trained inspectors.

If we bear in mind that these are proposals only, we can recognize the difference between European and American factory inspectors, and we can readily understand that important duties of the former, such as research work, cannot be performed by American factory inspectors but must be undertaken by other institutions, primarily the Divisions of Industrial Hygiene.

To care for safety and the control of health hazards is the task of other inspectors—if there is such a specification in the state. To train such other inspectors, the United States Department of Labor arranged training courses for state factory inspectors. The first of these courses was conducted in 1936, a ten-day lecture course that included visits to nineteen factories. An *Inspection Manual* (Bulletin Number 20) was published in 1938 as an aid to inspectors because "it is not to be expected that the average factory inspector is either a chemist or an industrial hygienist." In this way the Labor Department tries to improve the standard of the inspection.

Workmen's Compensation. It is apparent from the foregoing that modern industrial health is not a product of our workmen's compensation laws. These laws have, however, contributed substantially to its further development —perhaps more than in other countries, because here the laws followed much more quickly the first efforts toward industrial hygiene.

The federal government was the first to compensate for accidents to its employees (1908). Montana followed in 1909 with compensation for accidents to miners. By 1911, ten states had introduced compensation, and by the end of 1914, eleven other states. Today every state except Mississippi has a workmen's compensation law.

A forerunner of compensation for occupational diseases was the Act of Illinois of May 26, 1911, which attempted to regulate the use of poisons in working processes and provided that a right of action shall accrue to the injured person for any injury to health "caused by willful violation of this Act or willful failure to comply with any of its provisions." If the worker died, this right was extended also to his widow. An Act of June 10, 1911, gave compensation "for injuries sustained by any employee arising out of and in the course of the employment." The intention of the promoters of this law was to provide compensation for occupational diseases, but court decisions narrowed its effectiveness. Therefore more precise laws were published later.

In a few states, provisions for occupational diseases were included in laws passed later; in most states such provisions were added to Workmen's Compensation Laws by amendment, or were handled by special legislation. Compensation for occupational diseases was introduced in North Dakota in 1925, in Minnesota in 1927, in New York in 1930, and in the following years in other states—among them, West Virginia and Wisconsin in 1935, Rhode Island in 1936, and Delaware, Michigan, and Pennsylvania in 1937. Today compensation for occupational diseases exists in 33 states, the District of Columbia, and the four territories. There is general coverage of all occupational diseases in two federal laws (United States Federal Employees' Compensation Act and United States Longshoremen and Harbor Workers' Compensation Act, both published in 1934), in the District of Columbia, in eleven states and the four territories. Twenty-three states have published lists of those occupational diseases for which workers must be compensated. Most of these lists include many items.

The picture of the beneficial effects of compensation laws would be incomplete if we did not mention the activities of the great insurance companies. These companies send their safety inspectors to factories they insure, in order to study possible improvements in accident prevention and the control of industrial diseases. The advice of these inspectors is heeded because the quality of the safety and health installations determines the premiums which the individual factory must pay. The compensation for industrial diseases induced some insurance companies to engage such well-known industrial hygienists as A. J. Lanza and Warren A. Cook, and also to install laboratories for research in industrial hygiene. The larger insurance companies publish and distribute many pamphlets and booklets on industrial health and safety.

3

Organizations and Associations Cooperating in Industrial Hygiene

THE FORCES instrumental in forwarding the legislation and practice of industrial hygiene are many and varied. It is not possible to weigh the importance of one against the other, for the importance of each factor differs at various times. Therefore the order in which they are discussed cannot be considered indicative of their relative value.

We described previously the development of one of the most important factors, factory inspection. The important activity of factory inspectors, their research work, and their investigations which created the basis for new regulations and practices will be discussed in Chapter VIII. Just now we shall consider associations and organizations which work for better laws, more stringent enforcement of these laws, and better installations.

THE WORKERS' UNIONS

We have already mentioned the economic, cultural, and political rise of the working classes. We have mentioned also that certain of their organizations demanded measures from the legislatures. In addition to the principal demand for more hygienic working conditions, the demand for shorter working hours, and the repeated demands of miners for better ventilation, there was almost no hygienic question in which worker organizations did not intervene or cooperate occasionally. But it was not until the end of the 19th and the beginning of the 20th century that some of the workers' organizations initiated an intensive and long-continued struggle for the improvement of factory hygiene and especially for the removal or lessening of certain dangers. I wish to cite here only a few examples.

The painters of all countries, supported by physicians, asked that the use of white lead for house painting be prohibited. The French painters' union was particularly active. It arranged meetings in Paris, among them one presided over by the famous scientist Brouardel, at which another physician spoke on lead poisoning. Big posters with pictures of persons paralyzed and otherwise poisoned by lead were plastered on the walls of the houses. The *Assiette au Beurre*, a well-known comic paper of the time, devoted its entire issue of April 8, 1905, to the struggle against white lead, with caricatures of leading white lead manufacturers. The report of J. L. Breton to the Chambre des Députés on white lead (1907) comprises 660 pages with an appendix of 174 additional pages. The French law which forbade the use of white lead in the building trades was promulgated in 1909.

The German painters' union also waged war against the dangers of lead paints. In 1904 they published a booklet called *Der Kampf gegen die giftigen Blei farben* (The Struggle against Poisonous Lead Paints). In 1905 the German *Bundesrat* (Federal Council) issued a regulation which prohibited the dry grinding of paints and required that workers using lead paints be examined semiannually by a physician. This rule did not produce satisfactory results, and determined efforts for the prohibition of lead paints were continued. Between 1903 and 1922 the painters' union published eleven booklets, among them a lecture given by the Medical Inspector, Prof. F. Koelsch, at a general meeting of the union in 1921. Finally, in 1930, the use of white lead in painting the interior of buildings was prohibited.

In Austria the propaganda of the painters' union, supported by the health insurance fund of the painters and by certain scientists, achieved results after thorough research was conducted by the Ministry of Commerce, with Dr. I. Kaup as medical expert. A regulation of April 15, 1908, forbade the use of white lead in the interior of buildings.

Several German unions instructed one member of each of their executive committees to supervise all matters pertaining to industrial hygiene. They published books, pamphlets, and brochures on the subject. In 1911 the Union of Factory Workers, which was made up primarily of the workers of the chemical industry, published a book of 123 pages by H. Schneider on *Die Gefahren der Arbeit in der chemischen Industrie* (The Dangers of Work in the Chemical Industry), with tables and illustrations. Schneider's successor in the union, Gustav Haupt, received a medal from the German Society of Industrial Hygiene for his activity in this field. The central organizations of the unions and some individual unions sponsored lectures on the industrial hygiene of the various trades. These lectures were delivered by physicians (industrial hygienists) before shop committees or delegates in special courses, as well as in the great annual meetings. In 1926 the central organization of the Freie Gewerkschaften, the socialistic unions, appointed a physician as their consultant in questions of industrial hygiene.

On hygienic questions the work of unions and their collaboration with the factory inspectors, particularly medical factory inspectors, is of the greatest value because of their intimate knowledge of the circumstances existing in factories and of the health and welfare of their own members. For example, in 1911 the Union of Factory Workers asserted that lung cancer was more frequent among chromate workers than others. The research of hygienists revealed no corroborative evidence of any significance. The union repeated its assertion in 1926. Research conducted by a medical factory inspector brought to light no clear evidence but did reveal remarkable instances supporting the suspicion of the union. Fifteen years later the technique for the production of chromates was entirely changed by the industry. All dust was eliminated. In 1937, factory physicians of the chemical industry wrote a report on the exceptionally high incidence of cancer of the lung among chromate workers, observed in the last decades, and thus justified the original assertion of the union.[1] Lung cancer of chromate workers is now included in the German schedule of compensable occupational diseases. In similar fashion, other unions called the attention of medical factory inspectors to individual cases of poisoning, as well as to unfavorable conditions in single factories or in general.

The English trade unions were also very much interested in hygienic questions and in compensation for occupational diseases. They interceded on behalf of individual workers and also in matters concerning groups of workers, and asked for laws or regulations.

The earlier reports of factory inspectors treated the complaints more extensively and mentioned complaints of the unions reporting single violations of the laws. For example, the Lady Inspector in her report of 1905 shows that, of 1,167 complaints, 268 came from unions. The Chief Inspector's report of 1908 mentions that conferences between the inspectors and the secretaries of two important trade unions in London brought salutary results. Representatives of the unions served as members of departmental committees and commissions. The Royal Commission on Metalliferous Mines and Quarries (1910) included the secretaries of the National Union of Quarrymen and of two local unions. Representatives of the unions also testified as witnesses before committees composed of scientists and officials. The Committee on the Use of Lead in Potteries, for example,

[1] W. Alwens, E. E. Bauke, and W. Jonas, "Auffallende Häufung von Bronchialkrebs bei Arbeitern der chemischen Industrie, *Archiv für Gewerbepathologie*, VII (1937), 69–84.

heard as witnesses the General Secretary and three other officials of the Society of Pottery Workers, in addition to six members, other workers, and representatives of the employers. The Departmental Committee on Compensation for Injuries to Workmen examined approximately sixteen secretaries of unions.

When Thomas Legge retired from his post as Senior Medical Inspector of Factories, in 1927, the Trades Union Congress gave him the position of Medical Adviser to the Congress General Council.

The Trades Union Congress cooperates with committees on industrial hygiene and with the Ministry. The T.U.C. has a special committee to deal with workmen's compensation and factory legislation. There is regular contact with the Ministry of Labor concerning the Factory Act of 1937 and insurance. The Trades Union Congress of 1943 asked the Ministry for a regulation on the cleaning of upholstery materials and for more efficient protection of mothers before and after childbirth. The T.U.C. has been invited by the Minister of Labor to participate in a conference on industrial health and has been asked by the Royal College of Physicians to participate in a discussion on medical service in industry. In 1945, the Trades Union Congress asked for improvements in the canteen order, in the standards of lighting order, and others.

During the first decades of their existence, the unions of all European countries concerned themselves primarily with the questions of wages and working hours, drawing up some general principles of industrial hygiene but not going into details. Later, however, they developed a growing interest in questions of health, especially of industrial hygiene. The same development can be seen in the American unions. The most progressive of them have passed the first stage but, in contrast to the continental unions, their interest has broadened to include the whole field of public health. In England and Germany, public health, including health insurance, is the responsibility of state agencies or agencies instituted by the state, and it is left to the political party of the workers to see that their demands in this field are given serious consideration and implementation. In this country—lacking a workers' political party—it is the task of the unions to represent the workers in all questions of public health and welfare. Now (against strong opposition headed by the American Medical Association) American unions are fighting for passage of the Wagner-Murray-Dingell Bill, which calls for introduction of compulsory health insurance by federal law, as do the Priest and the Hill-Burton Bills. In some agreements between unions and management, a system of health insurance is provided. This activity in a larger field may be one of the reasons why only a few unions have developed an intensive activity in industrial hygiene, going into its details.

The interest of a union in industrial hygiene as early as 1911 was evidenced by a report on the investigation of New York sweatshops.[2] Following a needleworkers' strike, a joint board of employers and employees was set up to investigate and improve the deplorable workshop conditions. A sanitary standard for workshops was established. Eight inspectors—four men and four women—were appointed and a permanent organization developed. In 1912 a tuberculosis prevention campaign was initiated. Medical examinations were given to new employees, and those with infectious tuberculosis were excluded from work. A health center was set up for consultation and treatment of the members of the International Ladies Garment Workers Union. The inspectors ceased work in the year 1928 and inspectors of the state took over their work. But the health

[2] G. M. Price, *A General Survey of the Sanitary Conditions of the Shops in the Cloak Industry,* First Annual Report of the Joint Board of Sanitary Control in the Cloak, Suit, and Shirt Industry of Greater New York (New York, 1911).

center developed further and is today a great medical and dental institution of the Ladies Garment Workers Union. Similar centers of this union have been established in other cities.

Especially active in the field of industrial hygiene were and are the miners. The American Miners' Association (1861–1870)[3] asked political candidates to pledge themselves to support the miners' demands, among them a minimum health and safety standard enforced by public inspection of mines.

Paradoxically some wage agreements of the beginning of the century hindered rather than promoted the progress of safety measures.[4] These agreements expressly limited the duties of pit committees to the adjustment of grievances arising out of the agreement but did not permit the committees to discuss safety provisions. Some contained the stipulation that management of the mine and direction of the working forces are to be vested exclusively in the operators and that the miners will not abridge that right. The Illinois agreement of 1908 contained a section stating that the miners agreed not to "initiate or encourage the passage" of mining legislation "unless such proposed laws be mutually agreed to by the parties hereto or recommended by the Mining Investigation Commission appointed under the laws of the State of Illinois." Such a commission has existed in Illinois since 1909. It is composed of three representatives of the miners, three representatives of the operators, and three independent members not active in political life. In appointing them, the governor follows the wishes of the organizations. "In practice the legislature has never made any change in the mining laws that did not have the Commission's unanimous en-

---

[3] E. A. Wieck, *The American Miners' Association*, Russel Sage Foundation (New York, 1940).
[4] E. A. Wieck, *Preventing Fatal Explosions in Coal Mines* (New York, 1942).

dorsement. . . . Few changes of major importance have been made in the Illinois mining code in the past thirty years." [5]

In contrast to this, a wage agreement between District 10 of the United Mine Workers of America and the operators of the State of Washington since 1914 has provided for local joint safety committees composed of the president of the local union, the mine superintendent, and a third member selected by these two who shall not be a member of the pit committee. The safety committee shall investigate all serious accidents, make a bimonthly investigation of the mine, and submit recommendations to the manager or general superintendent of the company.

The Appalachian Agreement of 1941 has the following provisions: Reasonable rules and regulations of the operator for the protection of the persons of the mine workers and the preservation of property shall be complied with. At each mine there shall be a Safety Committee, designated by the district president of the United Mine Workers of America, consisting of a maximum of six mine workers not less than forty years of age and with fifteen years of experience. No member of the Mine Committee shall be a member of the Safety Committee. The committee shall have the right to inspect the mine. E. A. Wieck writes: "The beginning of collective bargaining on matters affecting safety is here . . . State mining codes and law enforcement agencies would not be displaced but supplemented." [6] The insertion of such clauses, requiring compliance by operators with state mining laws, subjects violators to discipline through the machinery of joint agreement or the operators' association, according to Wieck.

These local agreements were supplanted by the National Bituminous Wage Agreement of May 29, 1946, and the Anthracite Agreement of June 7, 1946. The first agreement,

[5] *Ibid.*, p. 109.   [6] *Ibid.*, p. 114.

between the Government Coal Mines Administrator and the United Mine Workers of America, covered the period of Government administration of the bituminous coal mines (a war measure) and had to be replaced by one between the operators and the union. The second was signed by the Anthracite Coal Operators and the union. The United Mine Workers of America (under John L. Lewis) were successful in obtaining these agreements after a strike. Besides regulations on wages and hours, three other demands were made: 1) a "Federal Mine Safety Code" for bituminous coal and lignite mines and "Federal Mine Safety Standards" for the anthracite mines, to be prepared and issued by the director of the Bureau of Mines after consultation with representatives of the United Mine Workers of America, the operators, and the Coal Mines Administrator; 2) mine safety committees, to be selected by the local union; and 3) periodical investigations by Federal mine inspectors who would report violations in the bituminous mines to the Federal Coal Administrator and in the anthracite mines to the management and the union.

The first application of the safety code was not very promising. After a big mine disaster in Illinois in March, 1947, the union accused the managers and the Coal Mines Administrator, to whom they were subordinate, of not cooperating sufficiently with the Federal Inspectors. An eight-day strike was called to give time for new inspections and to introduce improvements in all soft coal mines. As a consequence of the disaster and the poor results from the safety code, an agreement was reached between operators and the union on July 7, 1947; the code was incorporated in the contract, and an effort was made to strengthen enforcement. Wherever Federal Inspectors "find that there are violations of this Code and make recommendations for the elimination of such non-compliance, the Operators shall promptly comply with such recommendations"; either party to the contract who feels that compliance with the recommendations would cause irreparable damage or great injustice may appeal to a Joint Board of Review. There should be established a Joint Industry Safety Committee, two members appointed by the miners, two by the owners, to review and renew the code and to arbitrate any appeals. In each mine a safety committee, selected and paid by the local union, must inspect the mine and report to the manager. If the committee believes that imminent danger exists, the management shall, according to the committee's recommendation, remove all miners from the unsafe area.

It may here be pointed out that according to the law, codes of the U. S. Bureau of Mines have not the force of laws in the states and cannot be enforced by state inspectors; on the other hand, Federal Inspectors have by law no power to enforce anything.

The future must prove how the method will work. Enforcing safety and health measures exclusively by agreement between employers and employees and referring cases of non-compliance to the jurisdiction of a joint committee is in contradiction to the method in Europe, where it is the task of the state to supervise health and safety and to make and enforce regulations of this nature. The question arises whether such agreements can be enforced in times when the strength of the unions is weakened, as in a depression or in the gap between two agreements.

We shall return to this point later in discussing mine legislation (Chapter 12). It may be added here that many leaders of the American Federation of Labor back free enterprise: "Government interference is equally bad for both management and labor" (R. J. Watt). This may be explained by the dislike of many Americans for any governmental interference and by the discredit attached to the word "politics."

As a result of the strikes, the United Mine Workers won

a Welfare and Retirement Fund into which the managers must pay ten cents per ton of produced coal. The fund is to be administered by three persons, one representative of the employers, one of the union, and one neutral, chosen by the other two.

Other unions also are interested in factory hygiene. The United Automobile Workers (C.I.O.) suggested a "Proposed Health, Safety, Group Insurance and Compensation Clause" for inclusion in union contracts, asking, among other things, for health and safety committees.[7] The Health Institute of the U.A.W.–C.I.O. was established in Detroit, and in 1946 its Medical Diagnostic Clinic interviewed 3,411 patients and performed 8,298 laboratory tests and 3,555 X-ray examinations. The Health and Safety Education Department engaged a trained safety engineer, and conducted classes in health and safety (61 sessions) in cooperation with Wayne University and the University of Michigan.

Some agreements of other unions provide that sanitary and safety conditions are to be maintained in accordance with state, county, and city laws and that the company, in cooperation with the union, must work out a health and safety program. Other agreements state that the company "shall provide and maintain such safety and sanitation needs as are necessary to protect and preserve the health and welfare of all employees." Still others regulate the furnishing of goggles, rubber boots, and similar equipment. In some factories, workers must share this cost with the employer.

It is understandable that the agreements contain no detailed technical and other provisions, such as exist in the modern governmental regulations. The agreements are

[7] F. S. Mallette, "The Role of Industrial Hygiene in Labor Relations," *Industrial Medicine*, XIII (1944), 715. The C.I.O.—the Congress of Industrial Organizations—is one of the great organizations of unions. Above we mentioned the other, the A.F.L.—the American Federation of Labor.

written in general terms, with provisions similar to those of governmental rules of fifty years ago. Some agreements provide a work or safety committee, selected in part by the union.

Summarizing, we may say that European unions press for legislation and for the issuing and enforcement of regulations. Some American unions do likewise, but in consequence of the weakness and unreliability of the authorities in some states, other unions hope to bargain with the operators and, by their own strength, to make hygienic rules effective. In some cases such rules are formulated in part by government scientists. This procedure may be explained by the attitude toward political action mentioned above, but it is questionable whether it will prove helpful.

EMPLOYERS' ASSOCIATIONS

In every country there have been outstanding managers who promoted the hygiene of their own plants and tried to further the protection of workers in general by means of legislation; examples include the men who advanced the first protective laws in England, the work of Schuchard in Germany and that of John Buddle, English mineowner.

In all countries, discussions between the government and individual employers or employers' associations preceded the issuance of laws and regulations. In former times, the role of such associations was a retarding one. This changed slowly. Today the attitude of employers' associations is evidenced by the cooperation with the Home Office in England and by the work of the German Accident Insurance Associations, which are corporations of employers of single trades. In the United States large manufacturers and their associations cooperate in building up and financing national industrial hygiene associations, the importance and activity of which is discussed later.

INTERNATIONAL ASSOCIATIONS

As early as 1838 the Frenchman Blanqui asked for international agreements on labor protection. Villerme, another Frenchman, made the same request in 1840. Luc Le Grand, director of the Helvetic Republic, tried to achieve such agreements in 1840 and 1847. In 1866 the International Labor Association, at its meeting in Geneva, asked for international protection of the health of workers. In 1880 the Swiss *Bundesrat*, Emil Frey, requested that the Swiss Federal Council open negotiations with other states for international labor legislation. After the failure of the Berlin Conference of 1890 convoked by Wilhelm II, the Swiss Federal Council tried repeatedly to discuss labor questions with other states. International cooperation appeared urgent because, among other reasons, the belief was spreading that stronger labor protective laws weaken the power of competition of the more progressive states. The necessity for cooperation was also stressed by a congress of economists in Brussels in 1897.

International Association for Labor Legislation. After a Belgian committee had worked out a statute proposing such an organization, a Congress for Labor Legislation which met in Paris in 1900 decided to form an international association and, in connection with it, to create an International Labor Office. The duties of the Office were to collect a library, give information, publish a bulletin and prepare the meetings of the Association. The first meeting of the International Association for Labor Legislation took place in Basel in 1901. In 1908, a permanent Hygienic Council was created. In 1912 the Association had 7,011 members in fifteen national sections. Meetings of delegates were held every two years. At the last meeting, held in 1912, 118 delegates and 49 representatives of 19 governments were present.

The International Association for Labor Legislation gave strong impetus to investigations and legislation in industrial hygiene in all countries of the world. At its opening meeting (1901), three questions were selected to be dealt with first: nightwork of women, the manufacture and use of white lead, and the use of white phosphorus in the match industry. Stefan Bauer, the general secretary of the Association and of the Office, compiled the reports of the national sections on these questions into two books of more than 400 pages each, *Gesundheitsgefährliche Industrien* (Dangerous Trades), and *Gewerbliche Nachtarbeit der Frauen* (Nightwork of Women), Jena, 1903. In addition to these reports, the national branches of the Association conducted and published research studies in the individual countries, such as investigations in the phosphorus match industry in Austria (Teleky) and Hungary (Friedrich). The national sections carried on research on many other problems, among them (to cite only those on industrial hygiene) hand and machine typesetting, glazes, ferrosilicon, anthrax, hatmaking and hatters' fur cutting, caisson work, maximum weight to be lifted by longshoremen, and statistics of occupational morbidity and mortality.

In a competition, prizes were awarded to Richard Müller for his book, *Die Bekämpfung der Bleigefahr in Bleihütten* (The Control of Lead Danger in Smelting Works), Jena, 1908, and to M. Boulin for *Les Fonderies de plomb* (Lead Foundries), Paris, 1907. Leymann discussed other important results of the competition in *Die Bekämpfung der Bleigefahr in der Industrie* (The Control of Lead Danger in Industry), Jena, 1908. Theodore Sommerfeld and Richard Fischer drew up a *Liste der gewerblichen Gifte und anderer gefährlicher Substanzen, die in der Industrie Verwendung finden* (List of Industrial Poisons and Other Damaging Substances Used in Industry), which was published by the Hygienic Council of the Association (Jena, 1912).

Although only problems pertaining to factory hygiene have been mentioned here, the activity of the Association had a wider scope, including the work of women and of young people, working hours in mines, and other problems. The aim was to work out drafts of bills, based on exact research, which would be accepted and signed by as many states as possible. By 1912 the agreement on the prohibition of white phosphorus in the match industry had been signed by fourteen states.

League of Nations (International Labor Office). During World War I, the activity of both the Association and the Office had to be suspended. After the war, the League of Nations with its International Labor Office took over the activity, documents, and library of the former private institution. The International Labor Office continued to work in the same direction but on a larger scale, dealing with the most important questions of the protection of workers, such as the eight-hour day (which, by 1945, had been signed by 23 countries), employment of women before and after childbirth (signed by 16 countries), prohibition of nightwork of women (32),[8] prohibition of nightwork of young persons (32), weekly rest (33), sickness insurance (16), prohibition of underground work for women (22), and, in our special field, prohibition of white lead in painting of buildings (27), prohibition of nightwork in bakeries (12), and compensation for occupational diseases (14). Among the countries who signed these agreements were some whose signatures meant very little in this instance and whose implementation of the regulations remained questionable—Afghanistan, Albania, Cuba, Nicaragua, Iraq, and others.

Every year the International Labor Office called a conference of the member states, at which problems were dis-

---

[8] The figures in parentheses indicate the number of countries which had signed the agreement by 1945.

cussed and measures proposed, either in the form of "conventions" or "recommendations." The member states were obliged to bring the proposed conventions before the authority under whose jurisdiction the matter belonged within eighteen months.

The research work and publication of the section on industrial hygiene, headed by Carozzi, are likewise of great value. They deal with cancer of the bladder in aniline workers (1921), acetylene (1931), celluloid (1933), spray painting (1935), and other subjects. This section also supported the research on lung cancer among the Joachimsthal miners. It published the minutes of the International Silicosis Conference in Johannesburg (1930) and Geneva (1938) and of the International Conference of Medical Inspectors of Factories at Düsseldorf (1926). Highly important is the publication of the standard work, *Occupation and Health* (1925-1933), which was accomplished with the cooperation of many scientists from all countries. Supplements to this work are now being issued. Some of the publications of other sections also are worth mentioning, because they give a good survey of working conditions and institutions in all of the important and some of the less important countries. Among these are *Factory Inspection* (1923), *The Compensation of Accidents* (1925), *Statistical Methods for Measuring Occupational Morbidity and Mortality"* (1930), *The Compensation for Occupational Diseases* (1934), and a series on *Nutrition of Workers.*

NATIONAL ASSOCIATIONS

In addition to these international organizations, other associations have been active on a national scale. I mentioned above the national branches of the International Association for Labor Legislation, few of which survived World War I. The other national organizations are less concerned with furthering legislation than with spreading knowledge on industrial hygiene and answering practical questions.

In England, there was the Association of Certifying Surgeons; in Germany, the Institute of Industrial Hygiene. The latter was founded in the first years of the century by persons connected with the chemical industry. In 1910 it began publication of *Mitteilungen,* which was transformed in 1913 into the *Zentralblatt für Gewerbehygiene und Unfallverhütung* (Central Journal for Industrial Hygiene and Accident Prevention). This, however, had to be discontinued in July, 1922, in consequence of postwar difficulties. Soon after, the Deutsche Gesellschaft für Gewerbehygiene (German Society for Industrial Hygiene) was founded by representatives of the chemical industry, the unions, and scientists. From 1924 on, this society continued the *Zentralblatt für Gewerbehygiene und Unfallverhütung,* as well as special works written by experts, which the Institute had published. There were *Schriften* and *Beihefte* (additional issues), published up to World War II—altogether about 80 booklets, each dealing with a special problem in a specific way. The Society also arranged scientific lectures at the annual meetings, and conducted many courses in industrial hygiene for laymen, as well as for physicians and students.

In the United States, large organizations with ample funds play a very important role. In addition to research work, they are active in the field of "education"—the education of engineers, management, workers, and also physicians.

Initiated by the Association of Iron and Steel Electrical Engineers, the National Safety Council was organized in 1913 as a nonprofit organization. Its membership, now 6,700, is made up of industrial corporations, government departments, insurance companies, firms, and individuals. Every year the Association holds a National Safety Congress with an exposition. The records of the Congress are published. Since 1940 the Association has published an annual statistical review, *Accident Facts*. It also publishes eight monthly magazines of national distribution, including *National Safety News* (since 1919), *Public Safety* (since 1927), and *The Industrial Superviser*, and numerous pamphlets and posters.

In 1928 the American Standards Association developed from the American Engineering Standards Committee, which had been organized in 1918. The Association is concerned with technical problems, but as early as 1920 the same committee had established a Safety Code Correlating Committee, which published *American Safety Standard Codes*, sixty codes that helped to standardize governmental regulations. In 1938 it extended its activity by setting up "allowable" or "toxic" limits for dust and gases.

In consequence of the attention focused on silicosis in the United States in 1936, an Air Hygiene Foundation was created by some of the largest concerns (Allegheny Ludlum Steel, Aluminum Company of America, Du Pont de Nemours, Owen-Illinois Glass, and others), the Public Health Service, and the Bureau of Mines, in connection with the Mellon Institute in Pittsburgh. In 1941 the name was changed to Industrial Hygiene Foundation. The Foundation sponsored extensive research and published a "Medical Series" (on such topics as silicosis, X-rays of the lungs, and welding fumes), a "Preventive Engineering Series," and others. Most important is the *Industrial Hygiene Digest*, a monthly abstract journal of medical, engineering, and legal literature in industrial hygiene.

Also worthy of mention here is the American Foundation of Occupational Health, founded in 1915. Its present membership, totaling approximately 1,700, is composed of full-time factory physicians and those physicians who devote at least 50 percent of their time to work in plants. The Foundation attempts to promote medical training in industrial health and has established a clearing house for information relating to industrial medicine, hygiene, and traumatic surgery.

In his lecture on "Independent Agencies in the Field of Industrial Hygiene," delivered before the Annual Congress on Industrial Health in Chicago in 1939, McConnell enumerated other medical and technical associations, among them the American Society of Heating and Ventilating and the American Medical Association. Here we have restricted ourselves to those associations whose principal interest is industrial hygiene; otherwise we would find ourselves in an unlimited field.

4

Institutions in Factories for Promotion of Industrial Hygiene

SAFETY COMMITTEES AND WORKS COMMITTEES

England. In the annual report of the Chief Inspector of Factories for 1919 we read: "The experience of several British and American firms shows that apart from legal requirements, reduction of accidents can only be secured by gaining the interest and cooperation of operatives and officials through safety committees; but considering that the idea has been before the manufacturing community for several years it is somewhat disappointing to find that the setting up of such committees proceeds rather slowly." This institution did not find unanimous approval among employers. But the report mentions several very large factories which had excellent results with such committees. Some of the members were elected by the workers, some were appointed by the management. Monthly meetings were held, at which all the accidents were discussed and investigated by the committee. Two large firms also appointed safety inspectors.

There are also works committees, created to deal with all labor-management relations, some of which are concerned with safety. A committee or subcommittee discusses the accidents and makes tours of the plant to study possibilities for accident prevention.

The factory inspectors, attempting to improve safety education, suggested the organization of safety committees and the appointment of safety engineers. "But it must be admitted," the Chief Inspector noted in his report for 1926, "that their efforts have been attended with very moderate success and they have found it curiously difficult to arouse interest among either employers or workers." Consequently a Safety in Factories Order was drafted in 1927 for the iron and steel trade, shipbuilding, and some engineering and other industries. It was sent to the principal associations concerned to give them the opportunity to ascertain whether the employers would be prepared to take steps voluntarily to make arrangements of the kind contemplated in the order. The discussion caused the Secretary of State to hold up the order for the time being, to permit the organizations to take up the matter energetically with their members. A comprehensive "Safety First" campaign was launched, and numerous large firms installed safety committees or safety engineers, or both.

These safety organizations were further advanced by a resolution of the International Labor Conference in 1928 which stressed that the results obtained by legal regulations and state inspection should be improved and developed. The resolution pointed out further that regulations and inspection in themselves were not sufficient to prevent accidents due to "unsafe practice, fatigue, want of reasonable care on the part of the individual workers, lack of appreciation by new and especially young workers of the dangers." Therefore, steps should be taken along the lines of the "safety first" movement, all means should be employed to interest the workers in accident prevention, and safety organizations should be created in the plants.

During the following years, while the draft order mentioned was still in abeyance, safety organizations in England progressed. In 1929, of 1,129 factories which would be affected by the order, 849 had safety committees or other forms of safety organizations. Not all of them, however, were effective. By 1933, the number had increased to 982. There was a growing tendency in many concerns to replace the committees by a full-time safety officer.

In 1944, Chief Inspector Garrett wrote in his annual report: "The influence of Works Committees on accident prevention cannot be overestimated whether the actual work is carried out by the Works Committee or by a separate or Sub-Committee devoted to Health and Safety." Many committees existed in the best firms in prewar days; more than a thousand of these committees were in existence in 1937 and 1938. During the war their number increased. At the end of 1944, committees of one form or another dealing with aspects of work of special concern to factory inspectors were in existence in more than 3,000 factories. "Separate committees existed dealing with safety (943), health and welfare (652), canteens (1,904), while in addition there were 1,656 committees dealing with more than one of these subjects," according to the 1944 report of the Chief Inspector of Factories.

With reference to the miners, it may be mentioned that since the Coal Mines Act of 1872 the miners of every mine had held the right to nominate two of their number to inspect the mine. The Act of 1911 confirmed this right. The management was required to permit the inspection at least once a month.

Germany. The demand of the German workers for a works committee in every factory goes back to the year 1848 and the Frankfurter Parliament. According to a proposed bill, the committee should be composed of elected representatives of workers and management, to concern themselves with labor-management relations and to cooperate in drawing up work orders. The manufacturers at that time preferred factory works committees to unions. The Labor Protection Law of 1891 made workers' committees voluntary institutions in factories and gave them certain rights. The Prussian mine law of 1905 made them obligatory in mines, and required that they be consulted in drawing up work orders.

First a war law for certain war factories, then a law of

December 12, 1918 (published during the revolution), and finally the *Betriebsrätegesetz* (Law concerning Work Committees) of February 4, 1920, ordered that every plant with twenty or more workers must establish a workers' committee, to be elected by a secret ballot of the workers. In smaller workshops only one representative was elected. The task of the committee was to plead for the interests of the workers with the management. Its activities were very extensive. Section 66, Subdivision 8, of the law states that it was also the duty of the committee "to assist in the prevention of accidents and occupational illnesses, to support in this respect the factory inspectors and other such authorities by giving them suggestions, advice, and information, and to see to it that the rules laid down by the authorities and the instructions given by the insurance associations are complied with." Section 77 reads: "A member of the work committee so delegated by it must be present during investigations into accidents conducted in the plant, whether by the employer, the factory inspector, or other authorities." In larger plants, the committee designated one of its members to assist in accident prevention and care for the health of the workers.

As the law protected committee members from dismissal, they enjoyed a degree of independence from the management. In some larger plants, one or more members devoted their entire time to committee duties; they remained on full pay. Valuable services were performed by them in improving health installations and carrying through regulations. In addition to these committees, in several plants a worker was charged by the insurance association with the duty of assisting in accident prevention.

When National Socialism came to power, the relations between labor and management changed fundamentally. There was to be no difference of interests between them. Every plant was a "unit," the manager the "leader," the workers a body of "vassals" (Gefolgschaft). The leader was responsible for the well-being of his vassals. He had to nominate a *Vertrauensrat*, a committee of trustworthy men, and discuss with them all measures necessary for the improvement of factory hygiene. The "leader" could transfer to one of these men the task of dealing with all questions concerning the protection of workers. With this complete change in the conception of labor-management relations, the outlook regarding labor protection in the factory, factory hygiene, and the task of the manager and the duties of workers in relation to hygiene changed also.

United States. It is difficult to estimate the activities and importance of work committees and safety committees in the United States. In Public Health Service Bulletin No. 259, Bloomfield, Trasko, and others stated that, in the plants they investigated which had more than a hundred employees, 63 percent of the workers had shop committees (63 percent of the workers, not of the shops). In plants with less than a hundred employees, only 12.5 percent of the workers had shop committees.

The organization of these shop committees varies widely. In some plants, all members apparently are appointed by the management. In others, especially in mines, the members of the safety committee are appointed and paid by the union, in accordance with the Bituminous Coal Mines Agreement. In some agreements, members of the shop committee are expressly excluded from these safety committees.

The health and safety clauses proposed by the United Automobile Workers (C.I.O.) asked for a Health and Safety Committee to consist of three representatives of the company (including a safety official) and three representatives of the union. According to the proposal, union representatives may seek the advice of experts or authorities and have the right to call such persons into the plant, where they shall be permitted to make any necessary investigations. The committee is to be required to make weekly inspections and also such recommendations as they see fit at any time. We do not know how many such committees exist.

Safety committees with worker members who are elected by, and who have the direct confidence of, their co-workers seem to be infrequent. The effectiveness of safety committees varies widely, according to B. J. Stern. In *Medicine in Industry* (1946) he stated, "In some plants they are practically paper committees." In others they have the confidence and cooperation of both sides. More detailed statements are not obtainable.

SAFETY DIRECTORS AND ENGINEERS

The more complicated the technique of industrial hygiene and especially of accident prevention became, the more necessary it was to make an intensive study into its problems. Not every manager or engineer could devote sufficient time to this study in addition to his other responsibilities. Hence safety work became a specialty. In large plants, results were far from satisfactory when every department head met these problems in his own way. Therefore the larger plants delegated such tasks to especially trained men, who were made responsible for all safety and health devices of the big plant. This was especially true in large plants in the United States, where "safety directors" —full-time or part-time—have been introduced. The full-time work is particularly important; part-time may often be only a title.

According to Bloomfield and Trasko, 33 percent of the workers in plants with more than a hundred employees receive the "specified services" of a full-time safety director. But it must be remembered that 34 percent of these workers are in plants with over 1,000 employees, 16.8 percent of them in plants with more than 2,500.

FACTORY PHYSICIANS

The earliest forerunners of factory physicians seem to have been the mine physicians in Germany, unless there were such physicians in the Roman mines, which were worked by slaves. Among the most important German mine physicians were Georg Agricola, Theophrastus von Hohenheim, Samuel Stockhausen, and Martin Pansa. As the mines were a royal prerogative, the physicians were appointed by the government. They acted, as can be learned from their writings, as health officers, hygienists, and practicing physicians for the miners. Of similar mine physicians in England we could learn nothing, nor does the entirely different legal background of English mining make their existence probable.

England. Hyde reports that early in the 18th century in the Crowley undertakings a doctor, a clergyman, and a schoolmaster were retained voluntarily and jointly by the firm and the employees. The same author tells us that, according to a report by Edwin Chadwick in 1842, a Mr. J. Smith of Deanston, retained the services of a "medical gentleman to inspect the workers from time to time, to give them timely advice and as far as possible prevent disease." The factory inspection report of 1845 mentions a plant with 1,300 workers which had a surgery on the premises. The surgeon came daily between twelve and one o'clock. He also visited homes and provided medicine. He received a salary of £200 a year, toward which the workers who earned more than four shillings a week had to contribute one penny weekly.[1] We see here an early form of health insurance. Several similar health services were established later.

[1] R. Hyde, "Medical Services in Industry in Great Britain," *International Labor Review*, LI (1945), 433.

For a detailed discussion, see L. Teleky, "Certifying Surgeons—Examining Surgeons (1844–1944)" *Bulletin of the History of Medicine*, XVI (1944), 382.

FACTORY INSTITUTIONS FOR HYGIENE 101

We mentioned previously the tasks of certifying surgeons and the fact that, in some factories, surgeons appointed by the employer could make examinations, prescribed by law, if they had received the approval of the chief inspector of factories. Many certifying surgeons also seem to have concerned themselves with factory hygiene beyond their defined tasks as stipulated by law or regulations. E. R. A. Merewether, Senior Medical Inspector of Factories, states in his report for 1942 that the certifying surgeons "were often consulted by occupiers of factories and advised on matters of factory hygiene, First Aid and the like, and in many cases became in effect part-time works medical officers."

In World War I the Home Ministry prescribed a number of protective measures for the manufacture of trinitrotoluene. All factories which worked with trinitrotoluol had to be visited by the Health of Munitions Workers Committee. A physician was to be on duty in every plant at all times, and if there were more than two thousand workers, a second physician was necessary. Every worker had to be examined at least once a week. J. C. Bridge, Senior Medical Inspector, wrote in 1940: "Medical supervision in factories, although in existence, was greatly extended during the war of 1914–1919 and became universal where toxic substances were handled. From that time onwards such supervision in industry has been steadily, if somewhat slowly, developed." [2]

In 1931, the Industrial Welfare Society, founded by R. Hyde, called a meeting of the work doctors. Twenty work doctors and several certifying surgeons attended. Thomas Legge, formerly Senior Medical Inspector of Factories, now Medical Adviser of the Trade Union Council, pointed out at the meeting that the workers would not agree to a pre-examination paid for by the employer; they would agree

[2] *Annual Report of the Chief Inspector of Factories* (London, 1940).

only to a pre-examination performed by a part-time State Medical Service paid for by employers and employees in equal part.

It is perhaps a consequence of the increase of factory physicians (although such appointed physicians were mentioned in some earlier rules) that the Factory Law of 1937 states expressly that the Secretary of State "may authorise a medical practitioner who is employed by the occupier of the factory in connection with the medical supervision of persons employed in the factory, but is not otherwise interested in the factory, to act as examining surgeon for the factory for the purpose of examining and certifying the fitness of young persons."

In 1939 there were 300 factory physicians, 50 of them full-time. The figures for 1942 were 700 and 150, respectively. Among these factory physicians we find such well known men as R. E. Lane, now professor at Manchester University.

The Factory (Medical and Welfare Services) Order, issued by the Minister of Labour and National Service in July, 1940, empowered the Chief Inspector of Factories or any other inspector of factories to order every employer of the war industry to appoint as many full-time or part-time medical practitioners, nurses, and supervisory officers as the inspector deems necessary. In the annual report of the Chief Inspector for 1942, E. R. A. Merewether says: "No direction to appoint a doctor or nurse has in fact been given under the above-mentioned Order; the results achieved were themselves sufficient to recommend the services from factory to factory. In fact, the industrial health services, partial in operation as they are still, fill a gap in the Health Services of the country which it is inconceivable can be allowed to disappear when the immediate emergencies are past."

According to the 1945 report of the Social and Preventive Medicine Committee of the Royal College of Physicians, it is understandable that the question of an industrial health service is now considered to be connected with the vaster plan of a national health service; it should indeed form an integral part of the latter and should be administered centrally by the Ministry of Health. Just now, factory physicians are found primarily in big concerns employing a thousand or more workers. There are 175 full-time and 700 part-time factory physicians in England. "One of the immediate requirements of the people of this country is an industrial health service which includes provision for all who need it most—not merely the large and complex industrial concerns but also the small factory and workshop, the building industry, the transport services, offices, hotels and catering establishments." [3] According to the report, the aims of an industrial health service are:

a) To promote the general health of the worker by the provision of a good working environment and by fitting the the worker into that environment.
b) To prevent occupational diseases.
c) To assist in the prevention of injuries at work.
d) To organize and supervise a service for the emergency treatment and care of injured and sick workers at their place of work.
e) To take an active part in the restoration to full capacity of workers disabled by injury or disease; and resettlement of workers suffering from permanent disability.
f) To educate the workers in the preservation of health and promotion of well being.
g) To promote research and investigations.[4]

"Such an industrial health service," the report states, "should be staffed by medical inspectors, consultants in industrial medicine, whole-time and part-time officers together with the necessary non-medical personnel."

[3] "Second Interim Report of the Social and Preventive Medicine Committee, Royal College of Physicians, London, 1945." *British Journal of Industrial Medicine*, II (1945), 51.   [4] *Ibid.*

Another passage underlines the magnificent work done by the factory inspectors, both lay and medical, to improve standards of safety, health, and welfare in the factories and points to the necessity of cooperating with these inspectors. The report notes further that the financing of an industrial health service should not be done by industry, because the workers may be suspicious of medical officers paid by and therefore directly responsible to the management. In times of depression there may be the risk of a reduction of the service; small enterprises may be unable to afford the necessary payment. The Committee proposes, therefore, that this payment should form a part of the general "health budget" or should be met by a levy on output or profits.

It may be added that several authors have suggested the training of industrial physicians in special courses at the universities, several of which have indeed created a department of social medicine.

Summarizing, we may say that in England in the last decades, under the stimulus of two world wars, factory physicians have been installed in an increasing number of plants. Now we see the tendency in England to extend this development—but at the same time to lessen the dependence of these physicians on management and to include them in the general health organization, thus bringing them into cooperation with the factory inspectors.

Germany. In a later chapter we shall discuss the medical examination of workers in dangerous trades prescribed by rules and supervised by the government. In addition, some plants of heavy industry, such as smelting works, had a physician examine workers before they were employed for strenuous work. The chemical industry appointed physicians to supervise the health of the workers. Before the introduction of compulsory health insurance, in 1883, some plants had sickness insurance funds to which every worker of the factory had to contribute. The chemical industry, in particular, organized such sickness funds. Their physicians, appointed by the management, kept consultation hours in the factory itself. This was the case in the dye works at Höchst on the Main, for instance, from 1879 on. The German Health Insurance Act of 1883 and its later modifications gave the work sickness funds a special position in the organization of health insurance, with stronger influence and greater financial support from the employer than other forms of sickness funds received, and permitted compulsory health insurance to be provided by them.

Along with the rapid growth of the chemical industry, its sickness funds and the number of its factory physicians increased. The physicians formed an association which arranged meetings and prepared booklets such as leaflets for physicians concerning occupational poisonings, published in three editions (1913, 1925, and 1930), and instructions concerning the tasks of physicians under the law for compensation of occupational diseases (1925). At the "conferences," questions of industrial hygiene were discussed. At one such conference held during World War I, the leader of the factory physicians of the chemical industry, F. Curschmann, emphasized the higher susceptibility of women to the effects of certain toxic substances, especially benzol and its nitro and amido derivatives. Women therefore should be excluded from such work in peacetime, he stated; in wartime they might be employed, but only with greatest caution, and should be selected and supervised by physicians. Various improvements in chemical factories are indeed due to these physicians, and some of their number —Bachfeld, Floret, Schwerin, Gross, and others—published valuable contributions to industrial hygiene. But let us remember that in their publications they could write nothing that the management did not wish to see published. The carrying out of improvements proposed by factory physicians has always depended on the good will and understanding of management.

The part-time physicians—as I know from my own experience—did not have the same good training, and their activities were restricted to their special task, mainly the examination of certain endangered groups of workers. Generally they did nothing for the hygiene of the plant, and their clinical knowledge of the occupational disease in question was sometimes poor. Such part-time factory physicians worked almost exclusively in smaller plants of the chemical industry and in other dangerous trades where periodical examination of the workers was prescribed by rule. This examination was their principal task, and they fulfilled it not too conscientiously, a fact that was recognized when medical factory inspectors took over their supervision.

After World War I and the German revolution, matters changed. Medical factory inspectors were appointed in Prussia in 1921—Bavaria and Baden had had them much earlier—and the other states followed. The law required the medical inspectors to supervise the factory physicians and to approve their appointment, in so far as they were appointed to perform examinations prescribed by law. The position of all factory physicians was somewhat altered by the fact that there were now physicians in factory inspection who were experienced in this field.

When National Socialism came to power, it favored the appointment of factory physicians and assigned to them much of the responsibility for the health of workers, thus making the factory a public health center. Hebestreit, medical member of the highest council for public health of the National Socialist Party, pointed out that the methods in use to protect workers and to preserve human working power by laws and prohibitions are entirely correct, but need "to be complemented by a positive formation of the whole work process in regard to the protection of man and to the preservation of his efficiency." An increase of efficiency and a longer duration of the ability to work must be achieved by beautifying the work surroundings, by awakening joy in the work, by providing recreation for leisure time, in short, by another kind of living as a whole. All this must be promoted by guidance in healthful living. It is essential that factory physicians care for the health of the worker in his place of work. But despite these duties of factory physicians, the governmental supervision of factories must not be diminished.[5]

At a meeting of factory physicians on April 18, 1936, one of the leading physicians of the National Socialist Party, Dr. Bartels, explained that the factory physician should limit his treatment of the workers to emergency cases. He may give advice to other workers, but should not treat them. In addition to these duties in the factory, he must care for the life of the workers outside the factory—their sports, their exercise, their vacation activities.[6] In many plants, factory physicians have been assigned these additional duties. But the importance of factory inspection, especially the work of medical inspectors, is stressed in the publications.

Dr. E. Holstein stated in his inaugural lecture as associate professor at the University of Berlin, October 12, 1946, that the factory physicians under the National Socialist Party were in an unsatisfactory position, responsible both to the manager and to the party. They were required to submit very detailed records and statistics, which later

---

[5] Hermann Hebestreit, "Der Schutz der menschlichen Arbeitskraft im Rahmen der sozialpolitischen Neuordnung," *Zentralblatt für Gewerbehygiene,* XXIII (1936), 49-53.

[6] "Tagung der Betriebs- und Fabrikaerzte in Bad Nauheim am 18 April 1936," *Zentralblatt für Gewerbehygiene,* XXIII (1936), 107-108.

proved valueless. Holstein recommended that they should be made independent of the plant, but should be under the supervision of public health officials.

United States. Here conditions differ in several respects from those in Europe. In this country there are more big plants, and many of them are larger than the average big plants in Europe. There are many small workshops too. An exact comparison is not possible, because the figures are not easily comparable. But it may be mentioned that in 1937 the United States had 6,549 factories each employing more than 250 workers; among them were 241 factories, each employing more than 2,500 workers, or a combined total of 1,200,000. Moreover, factory inspection is inferior to that in Europe. Inspectors are not so well trained and are less efficient. There is also no compulsory health insurance.

For approximately ten years a movement has been under way for "industrial medicine"—principally for the appointment of factory physicians and industrial nurses, as well as for the care of the workers' health in the plants and under their auspices. The organizations most active in this movement are the American Medical Association, the American Association of Industrial Physicians, the National Association of Manufacturers, and the United States Public Health Service. How much has been accomplished it is difficult to say.

Extensive research was conducted by the Public Health Service into the problems of industrial health in the United States. The report made by J. J. Bloomfield, V. M. Trasko, and others presents a preliminary survey of these problems (Public Health Bulletin No. 259). Based on data submitted by 15 states, it covers the period from 1936 to 1939. Without going into detail, we must point out that an error appears to have been made in the data. The report indicates that at least 15 plants with less than 100 employees have a full-time physician, and at least 74 have a full-time safety director, which seems improbable. Statistical tables show the percentage of workers to whom certain services are available. The tables are divided by plants with more than 100 workers and those with less than 100 workers. In both groups, a first-aid kit is available for about 91 percent of the workers. In plants with more than 100 workers, a full-time physician is at the disposal of 20.3 percent of the workers, a full-time safety inspector is at the disposal of 33 percent, and a full-time nurse at the disposal of 43.7 percent. But we must keep in mind that 34 percent of these workers are employed in plants with more than 1,000 workers (nearly half of them in plants with more than 2,500) and a further 21 percent in plants with 500 to 1,000 workers. If the figures on physicians were given for the plants, grouped according to their size, rather than for the numbers of workers, the picture would be clearer.

In summarizing, we may say that provisions for "industrial medicine" (except first-aid kits) do not seem to be overwhelming. Some of the largest plants apparently have all possible facilities, and we may assume that much valuable work is done in these plants. But we would not be justified in assuming that this is true of the majority of the large plants—and certainly it is far less true of others. O. F. Hedley writes: "There has been too great a tendency for industrial hygiene physicians and engineers to gather at official conferences . . . and to discourse at great length what they are doing to protect the health of workers, while in reality their programs fall far short of these objectives. Sometimes the impression is quite evident that these mutual admiration contests are a sort of defence mechanism for what is being left undone." [7]

World War II brought great advances in these phases of

---

[7] O. F. Hedley, "Medical Services," in William M. Gafafer, ed., *Manual of Industrial Hygiene and Medical Service in War Industries* (Philadelphia, 1943), p. 43.

industrial hygiene. On the one hand, a great number of persons unaccustomed to factory work, particularly women, streamed into the war industries, making medical supervision more than ever necessary. On the other hand, huge factories were built in small villages and at remote distances from city centers. The management had to provide transportation and housing, as well as medical and hospital care. The big installations of the Kaiser plants may be mentioned as an example.

An outline of the functions of a medical service issued by the Division of Industrial Hygiene of the United States Public Health Service and a booklet entitled *Medicine in Industry*, compiled by B. J. Stern in 1946, envision a medical service which would be much more far-reaching than that of various British proposals. These two American plans go beyond the suggestions of the British Royal College in including "pre-employment examinations and periodic examinations," "treatment of plant injuries and occupational diseases," and "prevention and control of infectious diseases in the plant." The first of these we shall deal with later; the second interferes with the right of the worker to select his physician; the last goes a long way into the field of public health. The project as a whole would concentrate in the plant itself a great part of the customary public health program and of the care of healthy and sick workers. Unlike the British and other European projects and regulations, the American plans say nothing about cooperation with the factory inspectors and physicians of the Department of Labor, whose duty it is to enforce laws and rules protecting the workers' health. The British and German projects attempt to extend the authority of government inspectors and physicians, but the American plans are silent on that subject, although such an expansion may be especially necessary in this country.

It is apparent that the situation in the United States differs in many respects from that in Europe, not only because the governmental influence is weaker in this country, but also because there is no compulsory health insurance which would be—at least to some degree—replaced by provisions within the factory. Although some promoters of the American plans seek first of all to strengthen the management, others are far from such tendencies. In addition, some unions seek to wield a very strong influence over the health provisions and organizations in the factory. It is impossible, therefore, to predict the trend of future developments. But all sides desire an increase in the establishment of health institutions.

At the present time "industrial medicine" is widely used as a catchword. But it should not be forgotten that the field of the factory physician, as proposed in this country, would embrace the whole medical field now covered by any general practitioner, plus such specialties as certain occupational diseases and industrial hygiene, each only as far as they are of importance in the particular factory. It is not correct to speak, as E. H. Norris does, about "industrial surgery," "industrial ophthalmology," and "industrial radiology," [8] as if the accidents or illnesses of workers require a different kind of diagnostic or therapeutic procedure than accidents or illnesses acquired outside of factories. It may be added that, according to the 1946 edition of *Accident Facts*, published by the National Safety Council, 96,000 fatal accidents occurred in the United States in the year 1945; 45,000 of these occurred among workers, but only 16,000—or 17 percent—were occupational accidents. As these figures indicate, a small part of the fatal accidents are factory accidents.

[8] E. H. Norris, "A School of Industrial Health," *Industrial Medicine*, XI (1944), 780.

5
Methods of Preventing Injuries to Health

With the development of technology and with the growth of our knowledge in regard to dangers, regulations have had to be more detailed and more and more adapted to specific hazards. As an example of the ever more comprehensive and detailed hygienic requirements established by laws and regulations, we cited the paragraphs of British laws on ventilation and especially the rules concerning ventilation in grinders' shops. The same is true of every country: rules and regulations have become more numerous and their contents more precise, covering the details and peculiarities of every trade. Some of the principal methods of lessening health hazards, selected for and adjusted to the specific work, are prescribed in many rules and regulations. We cannot follow the historical development of each regulation here, but we shall make a survey of the most important of these methods and their historical development, dealing more extensively with those that seem to be of greater significance. We must stress again that the most important methods for improving health—shorter working hours, better nourishment in consequence of higher wages—cannot be treated here because they are beyond the scope of factory hygiene.

The methods developed for the control of health hazards (except accidents) may be divided in the following groups: the elimination of dangerous substances, the exclusion of especially endangered groups (women and children, and other groups by pre-examination and periodical examination), work turnover, shorter working hours (hygienic working day), personal precautions. and technical devices.

THE ELIMINATION OF DANGEROUS SUBSTANCES

The elimination of dangerous substances and their replacement by others may be accomplished by laws and regulations or by voluntary action on the part of manufacturers. The supposition of both is that science has discovered substitutes adequate for the special purposes of the dangerous processes.

We mentioned previously the prohibition of the use of white phosphorus in the match industry and of white lead in the painting of buildings, both ordered by law.

Matches, ignited by friction and using the poisonous yellow or white phosphorus, were invented in 1833 by J. F. Kammerer, who at that time was imprisoned as a German democrat. In 1847, Schrötter of Vienna discovered the nonpoisonous red phosphorus, which Böttger used in 1850 to invent the so-called Swedish matches, inflammable on a certain abrasive material only. The first cases of phosphornecrosis developed in the manufacture of matches were noted by Oberhofer in 1838 and described by Lorinser of Vienna in 1845. Other publications followed from different countries. An attempt was made to control the disease by regulations prescribing ventilation, periodical examination by a physician, and similar precautions. But the effect was poor. The invention of "Swedish" matches gave Finland (1872) and Denmark (1874) the opportunity to prohibit the production of yellow phosphorus matches. In their monopolistic factories, France introduced in 1898, Rumania in 1900 the fabrication of matches utilizing phosphorsesquisulfide and ceased to use yellow phosphorus. Germany prohibited the use of white phosphorus in the year 1907, Great Britain in 1908. In the United States, the Esch Law (1912) imposed a prohibitive tax on white phosphorus matches. We have mentioned the activity of the Association for Labor Legislation and the I.L.O. in this field.

Similarly, the use of lead paints could be restricted only after the discovery of paints fit to replace them. Courtois, of the laboratory of the Academy of Dijon, first made the proposal to replace lead carbonate by zinc oxide in 1780. Three years later, Guyton de Morveau of the same academy published the results of his experiments with various substances, which confirmed Courtois' proposal on zinc oxide; but its price was prohibitive. An Englishman, Atkinson, took out a patent for the use of zinc oxide in 1786. In 1835 a painter, Jean Leclaire, built a factory near Paris which produced zinc oxide for a moderate price. In 1849 a French decree ordered that zinc oxide be used in the construction of public buildings. Various ministries in France followed with different regulations. The production since 1875 of lithopone, a mixture of zinc sulfate and barium sulfate, offered another paint to replace white lead. We have reported the activity and success of the unions, the International Association for Labor Legislation, and the International Labor Office with regard to the use of white lead (see Chapter 3).

Mercury seems to have been introduced in the production of mirrors in about 1500 in Venice and two centuries later in Nuremberg and Fürth, Germany. In the middle of the 19th century there were many complaints, considerable research, and numerous regulations concerning the quicksilvering of mirrors. The method of replacing mercury by silver nitrate in the production of mirrors was invented by Drayton in 1843, and improved by Justus von Liebig, Petitjean, Böttger, and others. The first silver mirror factories were installed in 1856 in Belgium, by Petitjean; in 1857 in Switzerland, near Geneva; in 1858 in Erlangen, Germany; in 1859 near Nuremberg. In England, silver mirrors were produced preponderantly by 1870. The production of silver mirrors became cheaper than that of mercury mirrors when the price of silver went down. In the year 1889 the German government issued far-reaching rules for the use of mercury which made that process even more expensive. When I myself made investigations in Austria and Germany in 1911, I found mercury used in mirror covering rarely, in only a few factories, and temporarily for manufacturing mirrors for scientific purposes only.

Fire gilding, the dangers of which we mentioned earlier, has been replaced by the galvanoplastic method, invented by Jacobi, a Russian, and two Englishmen, Spencer and Jordan, in 1838. A large galvanoplastic factory was built in England in 1840 and 1841. Halfort, a German, does not mention galvanoplastics in 1845, but Pleischl writes in 1856 that fire gilding had almost disappeared. However, it has not disappeared entirely, but is still used, for example, in the gilding of steeples. Its main use is to gild military buttons, which must be made weather resistant; this use was noted by T. M. Legge in *Dangerous Trades* (1902) and by Beintker in the annual report of Prussian medical inspectors for 1933. The most recent cases of mercury poisoning from fire gilding in the United States were mentioned by Alice Hamilton in 1925. She states that there were three cases, according to the Civic Federation Report.[1] No such cases were mentioned in the annual report of the Chief Inspector of Factories (Great Britain) since the report of 1911, when one case was cited. During the last decades no cases of poisoning from this cause appear in the reports of the German factory inspectors.

The use of mercury nitrate in the felt hat industry, especially in fur cutting, was introduced in France in the 17th century as a secret process (hence the name "secretage" for this procedure). As early as 1815, the Paris Society for the Encouragement of Industry offered a prize for a method to replace mercury nitrate in the hat industry. Malard and

---

[1] Alice Hamilton, *Industrial Poisons in the United States* (New York, 1925), p. 249.

Desfossés applied for it but—it seems—without success. So, later, did Hillairet (1869). Many others followed. A method developed by Lussigne and Jourde in 1896 was introduced in the hat industry in Russian homes in 1910 but had no success in large factories. In 1926, E. Böhm and E. Elper described a new method which was tried in German factories in 1933. The United States Public Health Service, in consequence of its research into mercurialism in this industry, called a conference of employers and employees in Connecticut, according to Alice Hamilton. The conference unanimously passed a resolution prohibiting the use of mercurial carrot in the preparation of hatter's fur and the use of such carroted fur. The final outcome of all these experiments is not yet known.

The use of benzol in cement, which was greatly increased by the expansion of the rubber industry, has been forbidden in the manufacture of waterproof clothing in Belgium since 1936, and solutions containing more than one percent of aromatic products are prohibited in Belgium in rotogravure and mirror factories. Although other countries do not go so far, the substitution of toluol or xylol for benzol is recommended by governments and scientists, and has made progress.

As long as the use of dangerous substances cannot be avoided, other measures are necessary.

THE EXCLUSION OF ESPECIALLY ENDANGERED GROUPS FROM THE DANGEROUS WORK

Women and Children. The desire to exclude from work persons who are especially sensitive to the hazards of that work is at the root of all the laws that protect women and children and also prevent their employment in dangerous trades.

When an English law of 1833 prohibited the employment of children under nine years of age in textile factories and permitted children from nine to thirteen to work no more than nine hours a day, it became necessary —since reliable birth certificates were lacking—for a physician to testify that the individual was "of the ordinary strength and appearance" of a child of that age. This was, I think, the first medical pre-examination for factory work required by law. As it was impossible to get reliable certificates from ordinary practitioners, the factory inspector, Rickards, appointed "certifying surgeons," an institution which was made official by the Factory Act of 1844.

After the introduction of legal registry of birth, Inspector Robert Baker showed that certifying surgeons still were needed, not only because many birth certificates were falsified, but also owing to the fact that many children were rejected by the certifying surgeons by reason of bad health. So the certifying surgeons were retained. In 1901 they were given the right to issue certificates qualified by conditions as to the work on which the young person was fit to be employed. Thus pre-examination by a physician, first given as a substitute for the missing birth certificates, developed into a method to exclude physically unfit persons or others who might infect their co-workers. In 1944, 231,546 young persons between 14 and 16 years were examined. Of these, 3,578 were rejected—0.8 percent of the boys and 2.3 percent of the girls (the latter largely because of pediculosis). Certificates under conditions subject to nature of employment were given to 7,503; subject to re-examination were 3,511.

The law presumes that in certain kinds of work all women and children, by their peculiarities, are more endangered than are male or adult persons. Consequently women and children are forbidden to engage in such work. The British Mines and Collieries Act of 1842 prohibited the employment underground of all females and of males under ten years of age. The British Factory and Workshop

Act of 1878 prohibited the cleaning of moving machines by children, the employment of children and young persons in mercury mirror and white lead factories, and the work of young girls in certain parts of glassworks and brickworks. The regulation of May 7, 1898, prohibited the work of persons below the age of fifteen years in some divisions of the manufacture and decoration of earthenware and china. Sections 76 and 77 of the Factory and Workshop Act of 1901 likewise excluded women, young persons, or children from certain work.

Among the first German regulations of this kind may be mentioned the rule of May 13, 1884, regarding phosphorus match factories. Some German regulations of 1892 (glassworks, sugarworks, rolling mills, wiredrawers, chicory factories) and 1893 (brickworks) forbade the employment of women and young workers for certain functions. Later rules and regulations widened the number of activities from which women and young persons were excluded in both countries.

It may be noted that the exclusion of women from work with particular poisons came about because it has been definitely established—other assertions to the contrary—that women are more likely than men to be harmed by these poisons, as for example by lead, which damages their generative organs (Legge, Oliver, Teleky). Women also are more susceptible to such poisons as benzol and its derivatives because their blood-making organs are more sensitive to injuries (Curschmann). If there is a greater susceptibility of children to poisons, it has never been proved clearly for industrial poisons, though it seems probable. But it is certain that children are less careful than adults and therefore more endangered.

Other Persons Suspected of Being Especially Endangered. Exclusion from work—or from some types of work—also is indicated in the case of persons particularly sensitive to a substance because of certain bodily peculiarities and of persons who already have absorbed so much of a specific substance that curtailing the exposure seems necessary. The first group must be protected by medical pre-examination, the second group by periodical examination. These methods also have long been in use.

A regulation concerning factories making phosphorus matches, issued by the government of Lower Austria in 1846, made the following provision: "Such a factory must be visited by the health officer monthly and the workers, male and female, have to be examined, and, if one of them shows the slightest trace of a symptom attributable to the harmful influence of phosphorus, he must be eliminated from the work."

In 1853 the government of Middle Franconia, Bavaria, required that only workers whose sound state of health has been proved by the testimony of a legal medical officer shall be employed in the quicksilvering of mirrors. It required further that each worker be examined and instructed at quarterly intervals by this physician. Although this regulation justly required that the examinations be carried out by government physicians, a regulation issued by the Bavarian ministry on July 30, 1889, ordered the employer in mirror factories to have his workers examined fortnightly by a physician whom the employer appointed. The employer was required to communicate the name of the physician to the factory inspector. An exact registry of workers who were ill had to be kept. A rule issued in Berlin in 1888 on the manufacturing of incandescent lamps, in which mercury air pumps were used, prescribed weekly examinations by a physician.

The British rules concerning dangerous trades, which (on the basis of the Factory Act of 1891) were published by the British Secretary of State, extended protection by examination beyond that prescribed by the older British rules. A rule issued in 1892 required weekly visits to white lead factories by a physician, who had to examine every worker and enter the results in a register. The same applied to red lead plants, where the visits of the physician were to take place monthly. The regulation concerning the enameling of iron plates (1892) required the pre-examination of women and the monthly examination of all workers. No person was to be employed after an illness without the testimony of a physician. The regulation concerning potteries (1892) asked for monthly examinations of women and young persons, to be performed by the certifying surgeon. His findings were to be entered in a register, the form of which was prescribed by the Secretary of State. Such registers also are prescribed in the regulations concerning white lead factories (1899), but here the examination can be made by an "appointed surgeon," appointed by the manufacturer and approved by the Chief Inspector of Factories. The British regulations of the following years (on vitreous enameling, 1908; tinning of hollow ware, 1909; red lead, 1911; pottery, 1913, and others) required a complete health register. They also took another very important step forward, stating that pre-examinations and periodical examinations had to be performed in all the factories by the certifying surgeon.

A German rule of 1892 stated that young workers may be employed in glassworks only if they obtain a certificate of fitness from a physician. The German regulation concerning lead paint factories (1903) orders examinations to be made twice a month by a physician, whose name is communicated to the factory inspector and the health officer. The management of the factory must keep a control book which contains the names, ages, and other such data on all workers, the beginning and termination of their employment, and a record of every illness.

Germany, having no institution of certifying surgeons,

did not go so far as to prescribe such examinations through a health officer. Instead, "examination and supervision of the health of the workers by a physician approved by the governmental authority" is required by such regulations as those on nitro and amido compounds (1911) and lead paints (1920). After the introduction of medical inspectors of factories in Prussia (1921) and in other German states, these officials had to approve the appointment and supervise the activity of these physicians.

Thus we see that the trend of the development is to put the examinations as far as possible under governmental supervision. The reason is the same as that which caused the introduction of certifying surgeons in England: the impossibility of getting exact examinations and reports from physicians dependent on the manufacturer—or on the worker. I may add from my own experience that such examinations are reliably carried out only if efficient governmental supervision is exercised—perhaps with the exception of a few very large factories.

The value of the pre-examination of adults is that it enables us to prevent from entering dangerous work those whom we might expect to be harmed by the conditions of work in a higher degree than others, as a consequence of their physical or mental peculiarity. That is, persons with a deficiency innate or, in most cases, acquired through illness—for example, lung disease, which weakens the resistance against dust inhalation, or liver or kidney diseases, which hinder the excretion of a poison. That such persons are more endangered we know by practical experience.

A very different thing is the pre-examination for work that is neither dangerous nor especially heavy. There is no need to exclude from such work people who are not entirely healthy. Today these people have to make their living, and excluding them from large factories in which pre-examination is made means pushing them into smaller factories, often not so well equipped or protected, at perhaps a lower wage rate. On the other hand, selecting men for such work in a factory by medical examination is very difficult. In the words of V. K. Harvey and E. P. Luongo: "Examination of a human being to find what work he is fit for involves knowledge of a method not yet possessed by medical science, namely a method of estimating reserve compensation." [2] To be sure, pre-examination enables the management of a factory to select a healthy stock of workers and so to diminish absenteeism and the expenses of a sickness fund—but to do it in this way is no task of industrial hygiene.

Entirely different is the principle of the periodical examination in dangerous trades. Since the clinical symptoms of poisonings were thoroughly explored during the 19th and the beginning of this century, such periodical examinations are especially justified, theoretically as well as practically, in work which involves the danger of chronic poisoning, of silicosis, etc. The research of those decades showed that there are three stages in the intake of poisonous substances during a long period of time: (1) The poison is taken in and passes through the body, or is stored without any apparent signs or symptoms. (2) The effect of poison is visible. The poison produces some changes in the body, causes certain symptoms (such as lead line and light blood changes), but does not damage the function of any organ. At this stage we cannot speak about an illness or poisoning. Thomas Legge (1912) called this stage "absorption." (3) If there are signs or symptoms of the damaged function of an organ, then we must speak of an illness, of a poisoning. If the exposure is stopped after reaching a certain degree of the second stage, the outbreak of illness will be avoided. The advice of the physician (to cease exposure) must of

[2] V. K. Harvey and E. P. Luongo, "Physical Capacity for Work," *Occupational Medicine*, I (1946), 1.

course be executed by the management, as was required even in some of the oldest rules.

To halt exposure in time it is necessary that the intervals between the single examinations be equivalent to the degree of exposure; the rules provide different intervals for different trades. The examination must be performed by an independent, reliable, and experienced physician. It was the intention of all the newer rules to guarantee such a physician. Experience of this kind is more easily acquired by a physician who makes such examinations in a large district—as do the English certifying (examining) surgeons—than by a physician who sees only a small number of workers of one factory.

The periodic examination not only offers the possibility of protecting men from serious illness by excluding them from dangerous work at the appearance of certain signs. It also permits the discovery of the dangerous spots by observing the frequency or the quick development of the first signs among workers at a given place or work. Thus it controls the efficiency of prophylactic measures and indicates the necessity for their improvement. Finally such examinations, in conjunction with examinations of the air, provide us with the time of true "allowable limits."

An entirely separate question is the periodic examination of workers not exposed to specific dangers. This is a task of public health if not of the individual—and therefore should be undertaken by institutions of public health, in "health centres" such as are now provided for in the British National Health Act. They have also been discussed in this country. The examinations are of value only if they are continued over a period of many years. This cannot be done in a factory because the workers often change their jobs, but it could be handled effectively by a local center, which can transfer files to other such centers.

WORK TURNOVER

A much rougher method than the periodical examination for avoiding illnesses due to overlong exposure is the work turnover, which has no regard for the condition of the individual. After a certain time, all exposed persons are shifted to harmless work. After a period has been allowed for their recovery, they go back to the endangering work.

We find this method used in a very primitive way by the Spaniards who employed Indians in the first century after the discovery of America—and even today in the African colonies and in South Africa where native workers are employed. A still cruder method was usual in European white lead factories and zinc furnaces in the second half of the last century. For especially dangerous work—the emptying of white lead chambers or the cleaning of dust canals—workers were engaged from the road, often vagabonds. After doing this work, they were dismissed, sometimes ill.

The Viennese Collegium of Physicians, in an expert opinion of the year 1855, made the suggestion that women workers in phosphorus match factories should alternate their dangerous work weekly with other work. The turnover was introduced in the mirror industry in Paris at the beginning of the 19th century, and later in the Fürth mirror industry (Germany), in accordance with the regulation of 1853. The British rule of June 1, 1899, concerning white lead factories, provided that no one should work more than two days a week at emptying the white beds by the "Dutch process." In later times the turnover was used in other processes—for example, in World War I, in factories using trinitrotoluene.

SHORTER WORKING HOURS (HYGIENIC WORKING DAY)

We explained in the Preface to this book that we do not intend to deal with questions of working hours and their history, but working hours in dangerous or in certain very tiring occupations are a special problem. To the latter belong the "continuous industries"—blast furnaces, iron- and steelworks, glassworks—in which the International Association for Labor Legislation tried to bring about legislation for the replacement of the two twelve-hour shifts by three eight-hour shifts. These attempts met with partial success after World War I. To this chapter also belongs the problem of working hours in mines, especially mines with high temperatures.

Among the orders governing the "dangerous trades," one issued by the government of Middle Franconia, Bavaria, on May 25, 1837, was among the first to recognize the necessity of shortening the working day in mirror quicksilvering. This was also pointed out by Dr. Wolfring at the beginning of 1850, but was not made compulsory.

The British Factory and Workshop Act of 1901 gave the Secretary of State the right, in the regulations which he issued, to limit the hours of work, but this right was used only in the regulations for lead smelting and the manufacture of red or orange lead. The worker was forbidden to remain in a flue or condensing room longer than three hours without an interval of at least half an hour.

In 1902 the *Bundesrat* ordered a maximum work period of four hours for the vulcanization of rubber articles with carbon bisulphide or sulphur chloride. The German rule of 1906 prescribed a maximum working time of eight hours in the cleaning of furnaces in lead smelters, four hours in the cleaning of flue canals and rooms. The International Association for Labor Legislation discussed the "hygienic" working day in its meeting of 1910 and stressed the necessity of giving authorities the power to restrict working hours in dangerous trades. In the following two years the Bureau of the Association conducted research into legislation on this problem and reported to the meeting of 1912, which decided to work out a memorandum. Since that time more regulations of this kind have been published.

COOPERATION OF THE WORKER

In the older literature there were directions for personal care, most of them recommending certain medicines or foods. From the beginning of the 19th century on, the rules and regulations contained instructions on cleanliness, washing, bathing, and work clothes, and prohibited smoking and the taking of meals in the workrooms. Although the regulations were based in part on the overestimated importance of the intake of poison by the stomach and the skin, they were as a whole very valuable. In most cases, however, technical installations—especially those for ventilation—are far more important.

It was necessary to draw the attention of the worker to these rules of personal hygiene, in order to convince him of their necessity and to get his cooperation. As the first step to gain this interest and to teach him which protective measures were the responsibility of his employer and which ones he must himself observe, the earliest laws and regulations required that copies or excerpts of the laws be posted in conspicuous places in the factory. The British Health and Morals of Apprentices Act of 1802 contained such instructions, and subsequent acts and regulations usually had a similar clause. The Factory and Workshop Act of 1878 required that notices be exhibited showing, in addition to abstracts of the Act, the office address of the factory inspector and the certifying surgeon, the work hours, and the prohibition of employment of children in certain parts of the factory.

Under Section 11 of the Factory and Workshop Act of 1883, concerning white lead factories, copies of the rules, well printed in legible type, must be posted in conspicuous places; fines were specified for persons destroying such posters. The manager was required to give a printed copy of the regulations to every worker concerned who asked for it. All of the special rules issued by the Secretary of State at the beginning of the nineties had a chapter on duties of employed persons, which was to be brought to the attention of the workers.

In Germany, a regulation issued in 1853 by the government of Middle Franconia, Bavaria, on the manufacture of quicksilver mirrors ordered that the health officer who examined the workers quarterly must give them advice on personal hygiene. Most German regulations carry the requirement that copies of the rules be posted in the workrooms. In addition, the *Gewerbeordnung* (Trade Law) as amended in 1900 and 1908 provides that the management in every workshop with more than twenty workers must issue an *Arbeitsordnung* (working order) including information on working hours, wages, and similar matters. In some trades, such as lead paint factories, chromate factories, and printing shops, orders on hygiene also must be issued.

Experience has shown that the posting of such information is not sufficient to gain the cooperation of the individual worker. More important are repeated instructions, publicity by leaflets, the influence of the unions, and the efforts of work committees and other organizations devoted to hygiene (see Chapter 7).

TECHNICAL DEVICES AND THEIR CONTROL

The most important factors are always the technical installations. Regulations concerning these installations comprise the principal content of modern rules and must be carefully adapted to the specific processes and the peculiarities of the dangerous substances involved. Such technical measures are designed to eliminate poisonous or otherwise dangerous dust or gases from the air breathed by the workers. This may be accomplished by: (1) manufacturing endangering material in closed vessels and transporting it in a closed system of pipes, and (2) using efficient exhaust devices. The first is done in many industries, especially in the chemical industry. The breaking and crushing of dust-producing material in closed ball mills and disintegrators also have made great progress in this direction. We cannot go into the technical details here. However, all such apparatus must have openings for filling and emptying, and efficient exhaust installations must be attached to them to prevent poisonous gases and dust from reaching the respiratory tract of the workers except in quantities so small as to be innocuous.

All these technical devices must be tested and permanently controlled to determine the extent of their value in preventing dust or gases from contaminating the air. Twenty years ago the only method in use was a periodic examination of the workers. To this has now been added measurement of the amount of harmful substances in the air and the determination of their effect on health. For the development of industrial hygiene, it was necessary to find ways of measuring the noxious substances in the air. It was not sufficient to rely on general terms like "much" or "damaging," nor to infer from the incidence of disease in the factory that the air contained a damaging amount of dust or other substances. Only in the last decades has a fairly exact determination been possible, although attempts had been made since the middle of the last century.

Dust Sampling and Counting. We mentioned earlier the suggestion that the amount of dust in the air could be recognized by allowing the dust to settle. Among the attempts to measure the dust content of the air by this means are those recorded by Miquel (1879), Tissandier (1880), and Irvin (1902). In 1917, Edwin Higgins, A. J. Lanza, and F. B. Laney used this method in their research in mines. In

1925, K. B. Lehmann measured the dust which had fallen on a black piece of paper during specific periods in white lead factories. He also used other methods. In his *Lehrbuch der Arbeits- und Gewerbehygiene* (Textbook on Work and Trade Hygiene), Leipzig, 1919, he recommended that workers inhale through double nosepieces filled with cotton and exhale through the mouth, using a gasmeter. By weighing the dried cotton before and after the inhalation, the amount of dust contained in a certain amount of air and inhaled during a certain time could be calculated.

The method of drawing a measured amount of air through a filter by suction seemed rather simple. It was used by Hesse (1882), by C. LeN. Foster and J. S. Haldane in the air of mines (1905), and by I. Kaup in the research work of the Austrian Ministry of Commerce (1905). Kaup found that in a lead smelter in Carinthia there were, on an American stove with good ventilation, 0.2 to 0.3 mg. of lead (calculated as lead oxide) in 250 liters of air at the level of the worker's mouth. Kaup held that this was a negligible amount. Today we call 0.15 mg. in the cubic meter (1,000 liters) the permissible limit. In a white lead factory at Klagenfurt there were from 0.25 to 0.6 mg. of lead in 212 liters of air on the work places. At the opening of a crushing machine in a minium factory at Saag there were, despite exhaust installations, 1.4 mg. in 249 liters. In the breathing air of painters who were dry-grinding white lead paints, different tests showed 1.78, 8.85, 13.9, and even 25.01 mg. of lead in 1,000 liters (one cubic meter) of air. These figures explain the great number of cases of lead poisonings, especially among painters. In letter foundries, no lead was found above "complete type founding machines" (temperature 450° C.); but above a stereotype melting pot there were 2.5 to 9.0 mg. in 100 liters of air. The apparatus used for suction in these determinations was very primitive. It consisted of two bottles of 25-liter capacity each. One bottle, filled with water, stood on a higher level, the other on a lower level. The water, running from the upper to the lower bottle through a pipe which had a piece for aspiration, sucked in the air. The positions of the bottles were repeatedly exchanged. This apparatus was very cumbersome. So was that of Ascher (1909), which was built of metal sheets and contained a pair of two-piece bellows. In 1908 Martin Hahn constructed an apparatus with a small storage battery and an air pump which weighed, in all, seven kilograms (slightly more than fourteen pounds). This apparatus used different filtering materials: cotton, asbestos, nitrocellulose, and paper. In 1911 Brady and Touzalin constructed an apparatus consisting of a paper thimble which was filled with cotton. Although the used-air flow was very low, these filters were not efficient in retaining the dust. A more efficient filter was granulated sugar when used in greater amounts (100 grams), first applied in 1886 by Frankland (London) for a quantitative estimate of microorganisms present in the air.

Dust sampling devices were also constructed on entirely different principles. In 1914 Cave and his associates used a method that impinged a jet of dusty air on a sticky substance. In the preceding year E. V. Hill had produced a small apparatus employing the same principle. All this was done to examine the open air, especially in cities. Soon afterwards (1916), Kotzé's konimeter was constructed for the calculation of dust in the mines. (Kotzé was the chairman of the South African Miners' Phthisis Prevention Committee.) "The principle on which it is based is that of allowing the dusty air to impinge at a high velocity through a narrow nozzle against a glass plate coated with an adhesive substance." Several members of this committee and other authors recommended modifications of the konimeter, some of which have been incorporated into the "circular-konimeter," which allows many samples to be made without opening the instrument and without time-consuming manipulation.

As early as 1875, Coulier of France had shown that with dust particles as nuclei, a fog could be formed by suddenly reducing the air pressure and thus causing the water vapor of the air to condense upon them. In 1887 J. Aitken of Edinburgh constructed a "koniscope" based on his own experiments with this condensation method. Following the same principle, J. S. Owens built his "dust counter" in 1922, a very handy instrument which weighs only about three pounds. It is driven by hand and has been widely used in Great Britain and Germany.

Another method more often used for the absorption of gases is to make the air bubble through water in many bottles. On the same principle the apparatus of G. T. Palmer of New York was constructed in 1916. A measured amount of air is sucked through water, which simultaneously is sprayed through a flask.

In the year 1922 Leonard Greenburg and G. W. Smith, at that time employed in the United States Bureau of Mines, had the idea of combining the principle of collecting dust by impingement with the water washing or bubbling method. They constructed the "impinger," which is today the standard instrument in the United States. It can be driven by hand, by compressed air, or by electricity. The first models were very heavy. Now some modifications of the impinger in a handier form are in use, such as the midget impinger manufactured by the Mine Safety Appliances Company of Pittsburgh, which is driven by hand, has nine impinger tubes, and weighs slightly less than nine pounds boxed; the modification by Hatch, Warren, and Drinker; and others. In addition to these instruments there are still others, such as the electrical precipitator constructed by P. Drinker and R. M. Thomson in 1924, and less important ones.

Until now no instrument has been constructed which catches the dust completely. The impinger comes near this aim, catching 93 to 96 percent of the dust. Next to the impinger, the Owens "dust counter" seems to catch the greatest number of dust particles, especially the finest and therefore most dangerous ones, but not the coarser ones of more than five microns, because these do not pass through its small opening. The sugar tube, the Palmer apparatus, and the konimeter catch a smaller percentage of the dust.

All these apparatuses aim at taking the dust out of a measured amount of air as completely as possible. This having been done, the dust must be examined. In all the methods based on the principle of settling the dust, not even a fair estimate of the dust content of the air was possible. On the other hand, the methods using filters allowed only weighing. Since one particle with a diameter of 40 microns can weigh 37,000 times as much as one with a diameter of $\frac{1}{2}$–2 microns—and those smallest particles are the most dangerous—the weight is not important. What is important is the number of these small particles. They must be counted, which is possible only if the dust is collected through the modern apparatuses. In addition, the quartz content of the dust must be determined, because quartz is the most dangerous dust. Dust was first counted by means of an ordinary microscope with high magnification (480 times). Later the methods of dust counting were improved through projection and microphotography. The determination of the quartz content of dust is done through physical and chemical methods and X-ray diffraction.

Detection and Determination of Injurious Gases in the Air. In mines, small animals such as mice and birds have been used for centuries; their death indicated the presence of toxic or suffocating gases.

The first chemical examinations made in the mines or factories are believed to be the following. C. H. Brockmann reported in 1851 that Bodemann found in the air of the mine a few minutes after firing 1.80 volume percent less oxygen than normally, and 1.8042 volume percent and 2.7496 weight percent more of carbon dioxide. The figures for another mine were, averaging the results of from three to six analyses, 2.29 volume percent less of oxygen, and 2.3794 volume percent and 3.6262 weight percent more of carbon dioxide.[3] In 1899 Thorpe found, in the dipping room of a phosphorus match factory, 0.02 milligrams of phosphorus in 100 liters of air; in the boxing room, 0.12 milligrams. We referred above to the chemical examination of lead in the air, but that was for fumes and dust, not gases.

The lack of relatively simple methods and apparatus for determining the amount of noxious gases and the lack of knowledge as to the danger of the different concentrations of the various gases made it impossible to insert precise instructions on gases in rules and regulations for industrial hygiene. In the first two decades of this century, among the many European rules, only one contained a regulation concerning the chemical examination of the air. Even that one did not treat of harmful substances in the air but, as was customary, took the CO_2 content of the air as indicative of its quality. This was the English regulation of 1906 for the processes of spinning and weaving flax and tow. (The first regulation requiring the chemical examination of work material is the English regulation of 1913 "for the Manufacture and Decoration of Pottery," based on the method devised by Thorpe to determine the solubility of lead in glass.)

Of great importance in its consequences was Section 7 of the British Regulations for Chemical Works (1922): "Before any person enters . . . any absorber, boiler, cul-

---

[3] C. H. Brockmann, *Metallurgische Krankheiten des Oberharzes* (Osterode, 1851), p. 30.

vert . . . or other place where there is gas or fume, a responsible person . . . shall personally examine such place and shall certify in writing that such place . . . is free from danger." Then the question arose as to how such rooms could be tested. Mice are a reliable test only of the presence of carbon monoxide, for experiments have shown that they cannot always be depended upon to succumb to other important gases and vapors. There was a need for simple and rapid chemical or other methods to determine low concentrations of dangerous gases. The Association of British Chemical Manufacturers discussed the matter with the Home Office. Finally the Department of Scientific and Industrial Research, together with the Chemical Defence Research Department, developed a series of tests which were published under the title *Methods for the Detection of Toxic Gases in Industry*. The first tests, published in 1937, were for hydrogen sulphide; those for hydrogen cyanide vapors and sulphur dioxide followed in 1938; for benzol vapors, nitrous fumes, and chlorines, in 1939; and for organic halogen compounds, in 1940. It may be stressed that the "tests" dealt with are methods which permit the detection and measurement of the gas content of the air without the help of a laboratory. The method involves studying the effect of the contaminated air or a sampling of it during a given period on a filter paper impregnated with a certain chemical.

The Pertussi-Gastaldi test has been used since 1920 in the control of hydrocyanic acid as an insecticide and was soon prescribed by the German government. Later Koelsch and another German medical inspector of factories used this test in examining electroplating installations. In the *Zentralblatt für Gewerbehygiene, XXIII (1936), 177*, H. Weber gave a survey of such simple methods as using filter paper impregnated with a reagent, devised by himself (for arsine and phosphine), and by Smolczyk and Cobler and others.

PREVENTING INJURIES TO HEALTH 135

Another device for detecting and measuring small amounts of contamination in the air is the interferometer, "a delicate optical instrument which can be used to compare the refractive indices of two specimens of gas or liquid. . . . By comparing normal air with that encountered in an industrial environment, an estimate of its concentration of contaminants can be made." [4]

When we wished to ascertain the contamination of the air by a chemical determination, the difficulties which had to be overcome were basically the same as those concerning dust, but the practical solution of the problem was very different. The oldest method was to suck a certain amount of air through bottles filled with a liquid which reacted on the gas and fixed it. Formerly scientists, among them K. B. Lehmann (1919), believed that the air passing slowly—at a rate of six to eight liters per hour—through two wash bottles, connected in series, is cleared of its gaseous content. But it has been proved that in the usual washing bottles the substances in question are not completely retained. For example, air with lead fumes, when driven through wash bottles, shows traces of lead even in the seventh bottle. Therefore more complicated devices had to be constructed, following the principle of bringing a greater surface of air in contact with the liquid by dividing the air stream into many parts or even into bubbles. Such are the spiral and helical gas-washing bottles, constructed in different ways, as used by Weaver and Edwards (1915), by Friedrichs (1931) —whose apparatus was one of the most efficient—or the fritted glass-disk bubblers (1932). Instead of liquid, activated charcoal or silica gel may be used for some vapors. Then the gas or its compounds must be determined in the liquid or the absorbing substance. For different gases, dif-

[4] W. J. Burke, S. Moskowitz, and others, *Industrial Air Analysis, A Description of Some of the Chemical Methods Employed in the Laboratory Division of Industrial Hygiene, N.Y. State Dept. of Labor* (New York, n.d.).

ferent instruments and different liquids are necessary. The New York State Industrial Hygiene Laboratory standardized its absorbers for certain purposes and gave a detailed description of the methods used.[5]

All of these improved instruments and methods were constructed or designed in the last fifteen years, most of them since 1936. The finest methods were developed for determining the smallest amounts of various substances. The basic idea of one of these, the dithizon method for lead, was discovered by H. Fischer in 1925. Modifications were made in 1934. Later the method was developed and widely used. More sensitive methods for the quantitative determination of antimony were first described by Bamford in 1934 and by Anderson in 1939; those for cadmium, by Fairhall and Prodan in 1931. Those for nitrogen oxides were developed between 1932 and 1936. All these methods and also methods for the quantitative determination of combustible gases and for vapors of solvents, for benzol, carbon tetrachloride, chlorinated naphthalenes, methyl bromide, and many others were elaborated after the year 1929. American scientists are largely responsible for these developments.

It will be remembered that many of the poisonous substances now interesting us have been produced and used industrially only in the last fifteen years. That the exactness of the determinations is not complete and ideal is to be seen by the fact that the best of such methods always show differences in the examination of the same material. This means that in practice we must use a certain safety coefficient, which is necessary also in consequence of the differing sensitivity of different persons. We must agree with Jacobs, from whose standard work, *The Analytical Chemistry of Industrial Poisons* (1941), we took the examples mentioned above, when he says: "For the purpose of accuracy as far as the analysis of benzene in this instance is

[5] *Ibid.*

concerned, it is not significant whether 90 or 100 parts per million of benzol are present in the air, but rather whether 50, 100, 200, 300, 500, 1,000, etc., parts per million are present."

The simple tests of the English research institutes, which can be performed without the help of a laboratory, have been mentioned. In the last fifteen years, instruments have been developed which were based partially on such methods and tests for the detection and determination of contaminants in a quick and simple way. Such an instrument is the "carbon monoxide indicator," produced in the United States by the Mine Safety Appliances Company; it is graduated from 0 to 0.15 percent, and is produced in Germany by the Degea A.G. and the Draegerwerk. Among other devices are a benzol indicator which indicates from 0 to 1,000 parts per million and makes it possible to read concentrations directly to within 20 ppm, and a combustible gas indicator, both produced by the American firm named above; a "gas detector" for many different gases, made by Draeger; and the hydrogen sulphide detector developed by the United States Bureau of Mines, which shows hydrogen sulphide content between 0.0025 and 0.5 percent.

Determination of the amount of a certain substance in the air has its full practical value for industrial hygiene only when we know what effect this amount has in man. Many animal experiments have been made by toxicologists in the last decades. They are valuable for some purposes, but they do not allow any conclusion as to the effect on men of a specific amount of a poison, because there are wide differences in reaction between different kinds of animals. For example, guinea pigs show punctate erythrocytes after even minute doses of lead, while a goat which was given 320 grams of lead in ten months showed no symptoms. What we need are experiments or many experiences with men, and these as a matter of fact are not easily performed or acquired except in cases where the effect is only an irritating one. In all other cases, reliable statements—especially about chronic effects—can be obtained only by extensive observation of a great number of workers who are exposed to contaminated air. The state of health of the workers and the degree of contamination of the air to which they are exposed must be observed over a long period, at least some months, in order to find out whether or not there is the possibility of chronic effects.

It is only with a few poisons, those known for many years, that we have experience enough to fix the "allowable concentration" or "toxic limits," that is, the concentration which may be inhaled by men over a long period without damage. But even for such poisons the figures are not yet exact. For instance, the figures for benzol given by different authors are 50, 75, and 100 ppm. The inestimable value of such data, if they are exact, for working out means of prevention and elimination of damaging amounts of poisons has induced men to set up tables of such allowable limits. Provided they are used with caution and with an awareness of their unreliability, they can be very useful, especially if a safety coefficient is taken into account. But such tables are certainly out of place in rules and regulations (California, Connecticut, Oregon).

Warren A. Cook made a very valuable contribution not only by compiling tables of allowable limits from the rules of various American states (such tables are not used in Europe) and from official publications and literature, but also by giving the sources from which data concerning the various substances are taken, thus enabling the reader to check the reliability of the data to some degree.[6]

The use of such chemical methods in the construction and control of ventilation is widespread in the United

[6] Warren A. Cook, "Maximum Allowable Concentrations of Industrial Atmospheric Contaminants," *Industrial Medicine*, XIV (1945), 936–46.

States. They are not so much used in any other country, except Great Britain in a few cases. It should not be forgotten, however, that with every test we get only, as it were, a snapshot, and we do not know what happens before and after. We need therefore to have those tests repeated. In addition, we need a control to indicate whether the contamination of the air is permanently below the allowable limit in all sections of the workshop and how the workers react to it. The concentration of the air, changing often and differing very much, permanently influences the health of the worker and is in most cases to be recognized on examination of the workers.

Therefore a combination of the old method—regular periodical examination of the workers by a reliable physician—with the most modern method developed first by Americans—chemical analysis of the air, repeated in regular periods by a reliable chemist—may afford the best protection. This combination gradually will give us also the correct figures for "toxic limits."

Protective Devices Adapted to the Body: Dust Respirators. As it was not and will not be possible to use exhaust equipment in all places where men must work, there will continue to be a need for industrial masks.

We mentioned previously the alleged protection against dust and fumes provided by bladders or glass masks and a cloth bound over the mouth and nose in earlier times. There may also be mentioned the protective clothes and masks, often with peaklike mouthpieces, worn by physicians during plague epidemics in the beginning of the 18th century.[7] At the beginning of the 19th century, M. Gosse-Geneva recommended attaching to the mouth and nose a sponge saturated with a liquid which neutralizes the gas. M. Brizé-Fradin (1808) put a sponge

[7] John Gerlitt, "The Development of Quarantine," *Ciba Symposia,* II (1940), 567.

or cotton in a pipe-like tube, through which the worker breathed, the nose being left free. About 1825 a British miner, J. Robert, constructed a similar apparatus. For his invention he was honored with a silver medal, awarded by the British Royal Society of Arts. Later masks were fitted with layers of cotton wool (Tyndall), wool, or a piece of muslin. Patissier (1822) mentioned several types of respirators, but he stressed that all such respirators are a nuisance for the worker and therefore not to be used regularly.

In 1839 Tanquerel de Planches described a copper mask with glass pieces for the eyes and an opening filled with a wet sponge over the mouth. The mask was to be fitted tightly, allowing no air to enter at the sides. In 1836 Jeffrys of London constructed a leaf-like mask of finest gold and silver filaments. In 1849 Hazlett, an American, built a woolen filter which could be worn over either the mouth or the nose. A. E. Duchenne (Paris, 1861) describes *masques hygieniques* constructed by M. Paris. The facial part was of gutta-percha and fitted the face closely; in it was a complicated system with an inhalation- and an exhalation-valve. The filter surface was made of flannel, and the tendency was to make it as large as possible, giving the mask the form of a long, thick beak or a long tube. In 1905 a prize of £20 was offered in England for an "acceptable" dust respirator. Some sixty models and drawings were submitted, but none was considered suitable.

At the beginning of this century, the respirators on the market were inadequate in every respect. They could not be fitted to the face and had either too permeable filters or nearly impermeable ones. Slowly this improved. The first scientific research on dust respirators was published in 1911. The respirators examined by Schablowski allowed thirty to forty percent—and some of them even seventy to eighty percent—of the dust to pass through.[8] E. Brezina first recognized the importance of the "dead space."[9]

The South African Miners' Phthisis Prevention Committee tested respirators and came to the conclusion, as reported in the *General Report* (1916) and the *Final Report* (1919), that "Respirators have been found to be unsatisfactory."

Progress was made when that committee and government offices of other countries began to devote their attention to respirators and masks. By an act of Congress of February 25, 1913, the United States Bureau of Mines was authorized to test masks produced by manufacturers. It began its work about 1920 and intensified it after 1930. In 1921 the Safety in Mines Research Board in England and the *Preussische Grubensicherheitsamt* (Prussian Mine Safety Office) were established, both treating these questions. The Prussian office organized contests for dust protection in rock drilling, including a competition for dust masks.

Much research was necessary to determine the requirements for a good dust respirator. This research was done by Haldane, Simonelli, Dautrebande, Hesse, G. Lehmann, the American National Safety Council (H. F. Smyth), and the United States Bureau of Mines (S. H. Katz and others). Respirators, according to the reports on the research, should have a resistance to inhalation not greater than 24 to 36 mm. water for an air flow of 85 liters per minute and should be impermeable to the finest dust. Technique could not entirely fulfill the theoretical demands, but the conditions established in 1934 by the Bureau of Mines for the approval of respirators—resistance not exceeding 50 mm. water for an air flow of 85 liters per minute and a high

[8] Schablowski, "Ueber Respiratoren bei gewerblichen Staubarbeiten," *Zeitschrift für Hygiene und Infektionskrankheiten*, LXVIII (1911), 169–92.
[9] Ernst Brezina, "Ueber die Wirkung der gebräuchlichen Respiratoren," *Archiv für Hygiene*, LXXIV (1911), 143–63.

percentage of retention of finest dust—guarantee efficiency and relative comfort. In addition to bad dust respirators, there are now many good ones on the market, approved in the United States by the Bureau of Mines and in Germany by the *Deutschen Ausschuss für Staubschutzgeräte* (German Committee for Dust Protection). In England the Chief Factory Inspector orders respirators to be tested but communicates the result only to the factory inspectors.

Gas Masks. In earlier times the only difference between dust respirators and gas masks was that, for the purpose of retaining gases, the filtering material was soaked with a liquid presumably effective against a particular gas. Even until the beginning of World War I, "gas masks" consisted of a few layers of cotton or fabric impregnated with alkalis or acids, usually put between two pieces of wire gauze. Early in the war, the English and French armies used small packages of cotton, wool, or a similar substance, impregnated with chemicals, ordinarily a natriumthiosulphate solution (against chlorine). Later, in the Russian army, caps came into use which were worn over the head and tightly fixed around the neck by drawing together; their fabric was impregnated with various chemicals. The Germans had a mask which covered eyes, nose, and mouth and was connected with a container filled with a protective substance, through which the air for breathing had to pass.

After the war, especially in the United States but also in Germany, scientific and technical work on the construction of gas masks continued. In the resistance to breathing, the same figures are valid as for dust, but the Bureau of Mines found it necessary to concede a resistance up to 89 mm. water for an inspiratory air flow of 85 liter minutes in consequence of technical difficulties. The types commonly used today are "filter boxes" or "canisters." They are filled with porous granulated material which absorbs the chemical, reacting on the dangerous gases for which the mask is designed. For readily condensable vapors the filling is activated charcoal. Today in every country many kinds of masks exist, produced by various manufacturers. Each of these masks protects against one or a group of gases; a few of them, such as the American N-mask, protect against many gases. The protection is efficient only for the gases for which the mask is built and only against low concentrations and for a limited time.

A special problem was the protection against carbon monoxide. It was solved by the invention (1920) by J. C. W. Frazer and C. C. Scalione of a substance transforming CO to CO_2, a mixture of manganese dioxide and copper oxide called "hopcalite." The name is a contraction of Johns *Hop*kins University and the University of *Cali*fornia, in whose laboratories the research was done. German manufacturers used a similar substance. In the United States, some modifications are in use today.

Until now the British and the German governments have not allowed the use of gas masks in mines, for good reasons.

Hose Masks. Hose masks have their prototype in a certain apparatus recommended—and possibly used—for diving. Flavius Renatus Vegetius (1511) offers the picture of a leather bag, enclosing the head and entire body of the diver, attached to a leather pipe which reaches above the surface of the water. According to a description by the physician Chambon (1684), François Mercure van Helmont protected himself against gases in his chemical work by covering his nose with a handkerchief soaked in vinegar and breathing through a copper tube which was attached to his mouth and extended outside the room. In 1816 the Viennese fire brigade used bellows to blow fresh air through leather tubes into a container. From the container, the fresh air traveled through an inhalation valve and into a hood which covered the head of a man entering a gassy room. Today there are two kinds of hose masks: suction and pressure. The latter type, driven by hand or by machine, is especially useful for entering tanks and other closed rooms. Modifications are the "air line respirator" especially used in spray painting and the "abrasive blasting apparatus." These modifications, as well as the suction hose mask, are not effective in dangerous atmospheres.

Self-containing Oxygen Apparatuses. The first apparatus with a supply of air was constructed by Johann Alphons Borellus in 1682 as a diving apparatus. It consisted of an iron or tin vessel, two feet in diameter, that fitted over the head. Two pipes, closed by corks, could be opened to let out the used air and let in fresh air when the apparatus came from the water into the open air.

In 1799, A. von Humboldt suggested that one should carry an air supply when entering a room filled with irrespirable gases. In 1833, the Viennese fire brigade used a "rescue apparatus" that consisted of a large tin bottle filled with compressed air. In 1864, Galibert designed a "breathing bag" to be carried on the back.

The first regeneration apparatus containing chemicals and compressed oxygen was constructed by Schwann of Liége in 1854. A British naval engineer, Fleuss, built an apparatus to regenerate the exhaled air (1879). In Vienna, Gaertner and Walcher constructed the "pneumatophor" on similar principles (1895). After the production of steel bottles capable of resisting high pressure and of the Koerting injector, Bernhard Draeger of Lübeck built the first "circulation apparatus," which regenerated the exhaled air and added oxygen (1903). His apparatus was marketed over the whole world. Very soon other firms brought on the market other models based on the same principle.

Apparatus of this kind, called in the United States "self-contained oxygen breathing apparatus," in Great Britain "mine rescue apparatus," and in Germany *Kreislaufgerät* (circulation apparatus), must be constructed and handled with great care. Its main purpose is to make rescue work in mine disasters possible. If the rescue crew enters a section of a mine which is filled with irrespirable gases, every failure of the apparatus kills the carrier. Formerly such fatalities were exceedingly numerous. The first report of the British Mine Rescue Apparatus Research Committee (1918) begins with these words: "Owing to the unsatisfactory and sometimes fatal results obtained with some of the types of self-contained breathing apparatuses at present in use, we were appointed . . . to inquire into and to determine the advantages, limitations and defects of the several types. . . ." The United States Bureau of Mines and the Prussian mine administration authorities were active in the same direction. It is to their credit that today the inefficient types are excluded. In Great Britain and Germany, only apparatuses of a type approved by the mine authorities are permitted for use in mines. Today modern models of the Draeger apparatus (85 percent of the apparatus used in Germany are of this type), the Fleuss-Davis Protoapparatus, and that of Paul, Gibbs, and Mecco are in use.

The principle of all these apparatuses is the same: the exhaled air is freed from carbonic acid by a chemical (in the "alkali regenerator"); before this air is inhaled again, oxygen is added from a bottle containing compressed oxygen. The time during which the apparatus is efficient depends on the amount and efficiency of the chemical and—a simpler matter to ascertain—the amount of oxygen. We shall omit discussion on details of the different types, the various methods of regulating the air flow, whether breathing is done in a helmet or through a mouthpiece, and the danger signal before the complete exhaustion of the apparatus.

For use in British mines, a working time of two hours is required; in the United States all approved apparatuses, except one, must be capable of functioning for two hours; in Germany a working time of one hour is required. For factories, apparatus of a shorter working time is sufficient.

The apparatus must not only be well constructed; it must be kept clean, in good condition, and ready to use at any moment. It must also be tested at regular short intervals. In Great Britain this is done every month. The men who carry the equipment must be well instructed and trained—and retrained after short intervals. All this is an important phase of mine rescue work which has been developed in the last few decades. It will be discussed later.

Another type of apparatus contains chemicals which develop oxygen under the influence of the carbonic acid and the humidity of the exhaled air. The first apparatus of this type seems to have been constructed in 1902 by Degrez-Balthasard (France). Then came various types of the "Pneumatogen" apparatus invented in 1906 by Bamberger and Boeck of Vienna, the German "Proxylen" apparatus, and in the United States the "Servus" apparatus. For decades the experiments have been said to show "considerable promise of success," but until recently nothing of practical value has come out of them.

6

Prevention of Accidents

THE PREVENTION of accidents has proceeded steadily toward more clearly defined goals. The first laws of this nature required the fencing of movable parts of machines (British Factory and Workshop Act, 1878) and prohibited the cleaning of such parts by women and children. These fences, often painted red, are emblems of the period. Later there were attempts to make the machines themselves as safe as possible. The first regulations of this kind concerned steam boilers (Germany, *Bundesrats Verordnung,* 1890; England, Factory and Workshop Act, 1901). A very efficient change in construction was the installation of automatic apparatus for throwing a machine in and out of gear. Another such change was the replacement of square knife shafts with round shafts in surface planing machines; round shafts were in use in Germany by 1908 and were very soon made obligatory. Another such specification for the construction of machines is that every setscrew, bolt, or key on any revolving shaft, spindle, wheel, or pinion shall be sunk or otherwise effectively guarded (British Factory Act, 1937; Nevada Labor Law, 1937).

In the second decade of the century demands grew for a share in the responsibility to be given to the manufacturers of machines. It was requested that the manufacturer be obliged to build and sell only machines that conformed to certain safety laws and rules. We find such clauses in the recent laws (British Factory Act of 1937).

Employers in all times have combated protective measures and liability laws by arguing that the negligence of workers causes most of the accidents. Not until the beginning of this century did scientists begin to study the influence of factors other than technical installations on the frequency of accidents—Imbert in 1904, Pierracini and Maffei in 1907; since 1910, scientific publications have placed greater emphasis on the deficiency of workers as an important cause.

Everyone with practical experience knows that when the question is asked in an investigation, "To what extent did the fault of the worker contribute to the accident?" the final answer will be colored by the interest or viewpoint of the examiner. It is understandable, therefore, that the Bureau of Safety of the United States Steel Corporation in 1920 found that 90 percent of the accidents were caused by the inefficiency of the workers. The final report of the British Health of Munition Workers Committee of 1918, *Industrial Health and Efficiency,* quotes an American estimate that 10 percent of all industrial accidents are due to unavoidable causes, 30 percent to imperfections of machines, and 60 percent to apathy and lack of appreciation of danger on the part of operators. The British committee itself estimated this last cause at 25 to 40 percent. However great the fault of the worker himself, care must be taken to lessen it as far as possible. But the best basis for decreasing the number of accidents is the safest possible construction of all technical parts.

Beginning in this century, much research was done on the somatic and especially the psychic factors influencing accidents. In 1918, Vernon delivered to the British Health of Munition Workers Committee a memorandum entitled "An Investigation of the Factors Concerned in the Causation of Industrial Accidents." Other investigations were made by Vernon and also by Farmer in research carried on for the Industrial Fatigue Board. Many Americans worked in this field: Bogardus, Chaney, Kitson, Campbell, Goldmark, and Morgan. Among the Germans were Abelsdorf, Bienkowsky, Marbe, and O. Lippmann. Among the French were Imbert and Joteyko.

Research was conducted on the proneness of certain persons to accidents, sometimes in connection with the aptitude for the occupation; the influence of fatigue; the speed of work; the day of the week; state of nutrition; alcohol; sex; and age. Although some of the facts adduced could not easily be used in the prevention of accidents, all this research stressed the importance of the human factor in the origin of accidents—in contrast to the mechanical factor—and opened a new prospect for accident prevention and the field for the "Safety First" movement.

7

Education, Propaganda, Safety First

It was recognized many years ago that rules and advice and instruction by factory inspectors were not sufficient. A more far-reaching interest had to be created among employers, engineers, and workers by giving them further instructions (see Chapter 5, p. 125).

With this understanding as a basis, lecturers on industrial hygiene were appointed in several technical schools. Most effective in the spreading of knowledge were the exhibitions and the "museums" listed in Chapter 8.

It was in the United States, where "education" is one of the most frequently used catchwords, that a movement was launched to spread information on safety and industrial hygiene into the factory, to individual managers, engineers, and workers. It seems to have been initiated by the National Safety Council, founded in 1913. The activities of the Council increased constantly and embraced all groups of persons—managers, engineers, foremen, and workers. Here we shall concern ourselves only with the work done for the education of the last two groups. The National Safety Council now publishes approximately fifty different safety posters each month, a monthly magazine, *The Safety Worker,* and a magazine for foremen, *The Industrial Supervisor.* There are also "safety instruction cards"—more than five hundred pocket-size cards—for the various kinds of work, safety films, and "Safety in Foremanship" booklets. The Safety Speakers Bureau provides speakers for every kind of meeting. It may be added that the posters and pamphlets are not restricted to accident prevention but also deal in part with the prevention of industrial diseases.

Other organizations which we discussed earlier work in

the same field, as do the big insurance companies, particularly the Metropolitan Life Insurance Company. Federal departments and state departments also publish such booklets and pamphlets. This "education" is not confined to workers. The organizations and offices provide for the instruction of engineers, physicians, and management as well, and help them with advice.

It may be mentioned here that in the United States the element of education in our field is strongly stressed partly in conscious contrast to the issuing of rules and regulations, which is the very effective European way.

After World War I the idea of safety education programs, especially among workers, spread from the United States to Europe. The report of the British Chief Inspector of Factories for 1920 tells us that a very important safety conference was held in London the previous September. It was the first of its kind and was organized jointly by the Home Office and the British Industrial Safety First Association. Papers were read by representatives of five large firms. A paragraph in the report for 1920 reads: "Another interesting innovation worthy of mention was the 'staging' after the fashion which is so common in the United States of a 'Safety First' week at a large factory . . ." with posters, bulletins, and three-minute addresses to the workers.

Knowledge of the American methods understandably came to Germany much later. In the first annual set of the special edition of the *Reichsarbeitsblatt*, the *Arbeitsschutz*, 1925, are two articles about the American "Safety First" program, both exaggerating somewhat its wonderful efficiency but telling much that is interesting about the safety posters. There are also articles on the usefulness of pictures, with reproductions of American, Russian, Belgian, and Dutch posters. The Union of the German Accident Insurance Associations established in that year a nonprofit organization, *Unfallverhütungsbild, g.m.b.h.*

(Accident Prevention Picture, Inc.), and the governmental *Reichsarbeitsverwaltung* (German Department of Labor) published numerous such posters in 1925 and the following years. In 1929 an "Accident Prevention Week" was organized throughout all of Germany during the last week of February.

This propaganda made headway in all countries. It is certain that—combined with adequate technical control measures—the dissemination of safety information is successful. However, it is impossible to give very reliable statistical data, even though so many have been published, because it is extremely difficult to separate education from the complexity of contributing elements that work toward accident prevention.

8

Research and Studies

GOVERNMENTAL RESEARCH

RESEARCH AND STUDIES must precede all modern regulations. Factory inspectors during their visits to factories have the opportunity and the task of observing the conditions of work and workers, in order to discover the causes of accidents and diseases and to study methods for their prevention. In the first decades of factory inspection, its most important contribution—in addition to the enforcement of the laws protecting children and women—was in the field of accident prevention. This was the inspectors' responsibility as engineers, and most of the European factory inspectors were engineers.

The annual reports of the factory inspectors published in most of the European countries brought to light a tremendous number of observations and attempts to improve working conditions and eliminate certain dangers. The English, German, and Dutch reports are especially interesting; the French and Belgian also. It is deplorable that they are not studied more often by men working in this field, because observations of a permanently valuable character often are inserted in reports otherwise of merely local or temporary value.

Books dealing with accident prevention have been written by factory inspectors. One such is F. Reichel's *Die Sicherung von Leben und Gesundheit im Fabrik- und Gewerbebetriebe auf der Brüsseler Ausstellung, 1876* (Safety of Life and Health in Factory and Workshop at the Brüssels Exhibition, 1876) Berlin, 1877, a book of 84 pages. Factory inspectors collaborated on the following books: *Dangerous Trades*, edited by Thomas Oliver, London, 1902; *Handbuch der practischen Gewerbehygiene mit besonderer Berücksichtigung der Unfallverhütung* (Handbook of Practical Industrial Hygiene and Especially of the Prevention of Accidents) edited by H. Albrecht, Berlin, 1896; *Handbuch der Arbeiterwohlfahrt* (Handbook of Workers' Welfare) edited by O. Dammer, Stuttgart, 1902.

Other books treating of accident prevention and published in these years may also be mentioned here: J. Calder, *Prevention of Factory Accidents*, London, 1899; A. Pütsch, *Die Sicherung der Arbeiter gegen die Gefahren für Leben und Gesundheit im Fabrik-Betriebe* (Protection of Workers against Dangers to Life and Health in Factories) Berlin, 1883; M. Kraft, *Fabrikhygiene* (Factory Hygiene) Vienna, 1891.

If danger to health is caused not by accidents but by noxious substances—poisons and dust—or by temperature or overstraining, then more complicated studies and research are necessary which require specialized knowledge, especially that of a physician.

Only in England was research on the health of workers in different industries done to a great extent by the authorities as early as about the middle of the 19th century. Resulting from this research are the reports of the two great sanitary commissions of 1843-1845 and 1869-1871; the reports of the medical officers of the Privy Council, Dr. Simons and Dr. Greenhow in 1861, Dr. Bristons in 1862-1863; and the report of the Royal Commission on Metalliferous Mines, of 1864. Many other governmental investigations followed. Their results are set forth in such reports as that of the pottery committee (with the physicians J. T. Arlidge and W. B. Spanton) in 1893; the report of the departmental committee on workmen's compensation, 1904; the report of a committee appointed by the Admiralty to consider and report on deep-water diving, 1907; the report of the departmental committee on humidity and ventilation in cotton-weaving sheds, 1911; and the report of the Royal Commission on Metalliferous Mines (with the physician J. S. Haldane) in 1914.

In Germany another road was followed. Formerly health officers of the districts concerned were requested by the authorities to submit reports when complaints arose. Later, beginning in the 1880s, the *Kaiserliche Gesundheitsamt* (after 1918 called *Reichsgesundheitsamt*) was asked to conduct research and present opinions. This method continued up to World War II.

With the beginning of the 20th century, more and more attention was paid—without neglecting accident prevention—to the problem of "dangerous trades" and to other conditions undermining health. The workers, now more enlightened, became increasingly aware not only of the sudden and often dramatic accidents but also of chronically developing illnesses. The new chemical industry increased the number of industrial poisons used in many trades and even in small shops. The interest of ever-widening circles of physicians and factory inspectors was aroused by such illnesses. Even nonmedical factory inspectors studied these problems and published the results of their investigations. In Germany, H. Leymann wrote about lead poisoning in 1908 and 1914, R. Fischer about chromates in 1911. The growing importance of poisons and other harmful substances necessitated the appointment of physicians as factory inspectors in order to do specifically medical work. In previous times, physicians sometimes served as general factory inspectors and very often did excellent work—notably Horner, Baker, and Whitelegge in England, and Friedrich Schuler in Switzerland. But physicians were not appointed as factory inspectors with special medical tasks until later.

The first of them, with the years of their appointments, were:

Belgium	1895	D. Glibert
England	1898	T. M. Legge
Netherlands	1903	E. Wintgens, W. E. R. Kranenburg
Baden	1906	F. Holtzmann
Bavaria	1909	F. Koelsch
Italy	1912	G. Loriga
Austria	1919	Jenny Adler-Herzmark
Prussia	1921	L. Teleky, H. Gerbis
Saxony	1921	A. Thiele
France	1942	H. Desoille

The current observations and experiences of the medical inspectors and short excerpts from their more detailed investigations are published as a rule in the annual reports of the Chief Inspector, together with the observations of other factory inspectors. Usually special chapters are devoted to the activities of these physicians. In Prussia, separate annual reports of the medical inspectors of factories were published from 1921 to 1934. For the convenience of students, who found these reports difficult to consult because of the peculiarities of the material and the use of different languages, the reports were compiled, abstracted, and translated into German by E. J. Neisser for the year 1905 (Berlin, 1907) and later by E. Brezina and collaborators. These later volumes, covering the years 1909 to 1929, were published under the title *Internationale Uebersichten über Gewerbekrankheiten nach den Berichten der Gewerbeinspectionen der Kulturländer* (International Surveys on Occupational Diseases, According to the Reports of the Factory Inspectors of the Civilized Countries).

In addition to routine inspections and other current work, medical inspectors of all countries had to perform many thorough investigations, the results of which usually were published in official publications in Great Britain and in scientific journals or booklets in other countries. Only official studies which dealt with the most important special problems are listed below. Publications which the inspectors wrote as scientists, not on order of their governments, will be mentioned later.

The following works by medical factory inspectors are grouped by country.

ENGLAND

T. M. Legge. "The Health of Brassworkers," in the *Annual Report of the Chief Inspector*, 1905.
—— *Ulcerations of the Skin and Epitheliomatous Cancer in the Manufacture of Patent Fuel and of Grease*, 1910.
—— "Report on Body Temperature of Spinners and Weavers," Appendix IX of *Report of the Departmental Committee on Humidity in Flax Mills and Linen Factories*, 1914.
T. M. Legge, with A. M. Anderson and G. E. Duckering. *Special Report on Dangerous and Injurious Processes in the Coating of Metal with Lead*, 1907.
E. L. Collis. *Special Report on Dangerous or Injurious Processes in the Smelting of Material Containing Lead and in the Manufacture of Red and Orange Litharge and Flaked Litharge*, 1910.
E. L. Collis, with W. Sydney Smith. *Report on the Manufacture of Silica Bricks and other Refractory Materials used in Furnaces*, 1917.
E. L. Middleton and E. L. Macklin (engineer). *Report on the Grinding of Metals and Cleaning the Castings with Special References to the Effects of Dust Inhalation upon Workers*, 1923.
E. R. A. Merewether and C. W. Price (engineer). *Report on the Effects of Asbestos Dust on the Lungs and Dust Suppression in the Asbestos Industry*, 1930.

BELGIUM

D. Glibert. *Le Travail Industriel des peaux, des poils et des crins* (Industrial Work on Skins, Hair and Manes), 1921.

GERMANY

Franz Koelsch. "Gesundheitliche Erhebungen über das Malergewerbe" (Research on the Health of Painters), in *Erhe-*

bungen der Bayerischen Gewerbeaufsichtsbeamten über das Maler Gewerbe, 1912.

F. Koelsch. "Die Giftwirkung des Zyanamids" (The Poisoning Effect of Cyanamide), Zentralblatt für Gewerbehygiene, IV (1916), 113.

—— "Die Gesundheitsverhältnisse der Arbeiter in den Lumpensortir-betrieben" (The Health of Workers in Rag-sorting Workshops), Reichsarbeitsblatt, III: "Arbeitoschutz," 1930, p. 116.

—— "Untersuchungen über Staubgefährdung in Schamottefabriken" (Research on Dust Dangers in Chamotte Factories), Reichsarbeitsblatt, III: "Arbeitoschutz," 1932, p. 12.

L. Teleky, with J. Lochtkemper, Erika Rosenthal-Deussen, and Derdack. "Staubschädigungen und Staubgefährdung der Metallschleifer" (Dust Danger and Dust Damages to Metal Grinders), Arbeit und Gesundheit, IX, 1928.

L. Teleky with J. Lochtkemper. "Studien über Staublunge" (Studies on Pneumoconiosis), Archiv für Gewerbepathologie, III (1932), 418–70, 601–769.

L. Teleky, with Ilse Weikert. "Die Wirkung der Fabrikarbeit der Frau auf die Mutterschaft" (The Effects of Factory Work on Maternity of Women), Arbeit und Gesundheit, XIV, 1930.

L. Teleky, with Erna Zitzke. "Untersuchungen über das Bäkereczem und seine Ursachen" (Research on the Eczema of Bakers and Its Causes), Archiv für Gewerbepathologie, III (1932), 68.

In every country, factory inspectors have written many articles for journals and have delivered numerous lectures. Inspectors active in this work, in addition to those listed above, include Henry in England, Gerbis and Beintker in Germany, and Jenny Adler-Herzmark in Austria. The lectures delivered by E. L. Collis on industrial pneumoconiosis (Milroy Lectures, 1915), and by Thomas M. Legge in Boston (1919), and the reports made and lectures given at international congresses deserve particular mention. In addition to these official studies, many other books were written by factory inspectors, such as those by Legge and Koelsch. Some of these will be listed later.

In connection with the activities and publications of factory inspectors, those of other governmental or official corporations are worth mentioning.

In Great Britain during World War I, the Health of Munitions Workers Committee was set up. Among its members were such well-known physicians as George Newman, A. E. Boycott, L. E. Hill, and the medical inspector of factories, E. L. Collis. The committee published many "memoranda," including one on industrial diseases, one on ventilation and lighting, several on food, and an extensive report entitled *Industrial Efficiency and Fatigue*. To study the latter question, the Secretary of State for Home Affairs in January, 1917, asked the Department of Scientific and Industrial Research to appoint a committee. In cooperation with the Medical Research Committee (later called the Medical Research Council), an Industrial Fatigue Research Board was created. Through its investigators, this board published 54 reports on research conducted between 1919 and 1929. In accordance with the original plan, the first studies concerned fatigue and efficiency as influenced by various conditions (hours of work, season, atmospheric conditions, lighting, monotony). Other reports covered vocational guidance, accidents, and morbidity. Among the more important of these investigators were H. M. Vernon, S. Wyatt, H. C. Weston, J. A. Fraser, E. Farmer, and Major Greenwood. Later the duties of this board were extended, and the Industrial Health Research Board was founded as part of the Medical Research Council. A total of 89 reports, including the 54 previously mentioned, some short, wartime "emergency reports," and 17 "war memoranda" have been published.

An interesting example of cooperation between agencies is the following, initiated during World War I. In August,

1915, the medical inspectors of factories, Legge and Collis, asked the Applied Physiology Department of the Medical Research Committee for assistance in an investigation into the toxicology of trinitrotoluene, which had caused the first fatality in the industry in February, 1915. Now began a joint operation of different organizations to study and clear the new poison. In October, 1916, the Ministry of Munitions published instructions for factory physicians. The Royal Society of Medicine discussed this problem in January, 1917, and published its proceedings. On March 30, 1917, a regulation was issued by the Ministry. The National Health Insurance Medical Research Committee in 1917 published *The Causation and Prevention of Trinitrotoluene (T.N.T.) Poisoning*. But the research work went further, and in 1921 the Medical Research Council published the results of experiments on animals. Another fine example of coordinated efforts against an industrial danger is that of the investigations of mule-spinners' cancer.

The German Health Board and the German Health Office (*Reichsgesundheitsrat und Reichsgesundheitsamt;* before 1918 these were *Kaiserlich- instead of Reich-*) had a special committee and a special department of industrial hygiene, the head of which, for the last two decades, was H. Engel. The results of the research were published in the *Arbeiten aus dem Reichsgesundheitsamt* and also in various journals. Between the years 1889 and 1938, among several hundred investigations on public health 46 on industrial hygiene were published. Some dealt with the various lead trades, others with mercury (1889), anthrax (1902, 1907, 1914, 1917, and 1933), Thomas slag (1899), and ankylostomiasis (1906). Among the most important reports of permanent value seem to be those of Löbker and H. Bruns on ankylostomiasis (1906); K. Beck and P. Stegmüller on the solubility of lead compounds (1910); H. Engel on injuries to health in working with metallic lead, especially in soldering (1925); Engel and V. Froboese on lead volatilization in lead-burning with different flames (1925); H. H. Weber on the analysis of technical solvents (1931, 1933, 1936), and the same author on the determination of poisonous gases and vapors in the air (1936). Another important work was the "Instruction for Medical Examination of Lead Workers" (1930), prepared by the *Reichsgesundheitsamt*.

In Austria, the Minister of Commerce ordered investigations of lead smelters, the manufacture and use of lead paints, and other lead trades. These investigations were conducted between 1904 and 1915 by Dr. I. Kaup, and the results were published.

In the United States, factory inspectors as a rule are less highly qualified and less carefully selected than in England and Germany. American factory inspectors therefore cannot be expected to carry on research and scientific work as do the inspectors in these European countries.

But the Federal government and several of the states have created other administrative bodies which could fulfill this purpose. Federal offices are the Division of Industrial Hygiene of the United States Public Health Service, the National Bureau of Standards, the Bureau of Mines, and various divisions of the Department of Labor. The states have their Departments of Labor and several have also a Division of Industrial Hygiene. That Federal and state authorities initiated special investigations sometimes even before the creation of special departments has been noted. There are so many departments and bureaus, each with a large staff doing research work in our field, that it is very difficult to select from the great number of their publications those which merit special mention. A further complication is the fact that usually the names of so many cooperating scientists are listed on the title page that it is difficult to ascertain who did the principal work.

Since 1915 the United States Public Health Service has published reports on 76 major investigations in industrial hygiene. Among them are those on dusty trades, lead trades, carbon monoxide, and aromatic amino and nitro compounds. Some of the most important authors were R. R. Sayers, W. C. Dreessen, and L. Greenburg, surgeons; Louis Schwartz, dermatologist; J. M. Dallavalle and J. J. Bloomfield, engineers; D. K. Brundage and R. H. Britton, statisticians; S. H. Webster, chemist; and L. T. Fairhall and W. F. von Oettingen, toxicologists.

The following Public Health Service publications, grouped by subject, have been selected with regard to the importance of the question treated. Unfortunately this necessitated the omission of other valuable works.

Silicosis and Dust

A. J. Lanza and S. B. Childs. *Miners' Consumption* (Bulletin No. 85), 1917.

L. Greenburg and C. F. A. Winslow. *Dust Hazard in the Wet and Dry Grinding Shops of an Ax Factory* (Report No. 35), 1920.

L. Greenburg, S. H. Katz, G. W. Smith and others. *Comparative Tests of Instruments for Determining Atmospheric Dusts* (Bulletin No. 144), 1925.

L. R. Thompson, D. K. Brundage, A. E. Russell, and J. J. Bloomfield. *The Health of Workers in Dusty Trades* (Bulletin Nos. 176, 187, 208), 1928, 1929, 1933.

J. J. Bloomfield, J. M. Dallavalle, and others. *Anthracosilicosis* (Bulletin No. 221), 1936.

W. C. Dreessen, J. M. Dallavalle, and others. *Pneumoconiosis among Mica and Pegmatite Workers* (Bulletin No. 250), 1940.

R. J. Flinn, J. L. Jones, and others. *Soft Coal Miners* (Bulletin No. 270), 1941.

Poisons

P. A. Neal, R. H. Flinn, and others. *Chronic Manganese Poisoning in an Ore Crushing Mill* (Bulletin No. 24), 1910.

P. A. Neal, R. R. Jones, and others. *Mercurialism in Hat and Fur Cutting Industry,* Bulletin No. 234 (1937), Bulletin No. 263 (1941).

W. F. von Oettingen. *The Aromatic Amino and Nitro Compounds* (Bulletin No. 271), 1941.
—— *The Aliphatic Alcohols* (Bulletin No. 281), 1943.
—— *Carbon Monoxide, Its Hazards* (Bulletin No. 290), 1944.
D. D. Donahue and R. K. Snyder. *Trinitrotoluene* (Bulletin No. 285), 1944.

SKIN

Louis Schwartz. *Skin Hazards in American Industries* (Bulletin Nos. 215, 229), 1934, 1936.

The Bureau of Mines works in many fields: mineralogical, geological, chemical, technical, mercantile, and hygienic. It has published numerous bulletins, technical papers, economic papers, schedules, miners' circulars, annual reports, and the *Minerals Yearbook*—in all, nearly 500 bulletins and 3,700 reports. Up to the end of 1946 the Health Division had published four bulletins, 20 technical papers, three monographs, five miners' circulars, 112 reports of investigations, 46 information circulars, 116 articles in journals, and 42 publications in cooperation with the United States Public Health Service. Among the most important scientists are R. R. Sayers, D. Harrington, Sara I. Davenport, H. H. Schrenk, and W. P. Yant. We mention here only a few publications:

E. Levy. *Compressed-air Illness . . . at the East River Tunnels*, 1922.
A. C. Fieldner, S. H. Katz, H. V. Frevert, and E. G. Meiter. *Gasmasks for Protection . . . against All Gases*, 1925.
S. H. Katz, G. W. Smith, and E. G. Meiter. *Dust Respirators, Their Construction and Filtering Efficiency*, 1926.
W. P. Yant, H. H. Schrenk, and others. *Urine Sulphate Determination as a Measure of Benzol Exposure*, 1936.
H. H. Schrenk. *Testing and Design of Respiratory Protective Devices*, 1939.
W. P. Yant, R. R. Sayers, E. Levy, and others. *Carbon Monoxide in the Air of the Holland Tunnel*, 1941.
G. W. Grove. *Self-contained Oxygen Breathing Apparatus*, 1941.

J. J. Forbes, M. J. Ankeny, and Francis Feehan. *Coal Miners' Safety Manual*, 1942.

In addition, there are publications on the midget impinger and on dust sampling and counting by J. B. Littlefield, F. L. Feicht, H. H. Schrenk, and others (1937, 1938, 1940, 1944); on quantitative analysis of dust by X-ray diffraction by J. W. Ballard, H. J. Oshry, and H. H. Schrenk (1940, 1942, 1946); and on microcolorimetric determination of benzene by Schrenk and S. J. Pearce (1935, 1936). Experimental mine investigations conducted by the Bureau of Mines were described by G. S. Rice (1911–1932).

There are also many publications from the United States Bureau of Labor and its different bureaus, the Women's Bureau, the Children's Bureau, and the Bureau of Labor Statistics. We may mention here several studies by Alice Hamilton which were published in bulletins of the Bureau of Labor Statistics: *Lead Poisoning in the Smelting and Refining of Lead* (Bulletin No. 141, 1914); *Industrial Poisons Used or Produced in the Manufacture of Explosives* (Bulletin No. 219, 1917); *Effect of Airhammer on the Hands of Stonecutters* (Bulletin No. 236, 1918); *Industrial Poisonings in Making Coal Tar Dyes* (Bulletin No. 280, 1921); and *Occupational Poisonings in the Viscose Rayon Industry* (Bulletin No. 34, 1940). Her *White Lead Industry in the United States*, published in 1912, before the Bureau of Labor Statistics was organized, contains a valuable appendix on the lead oxide industry.

The Division of Industrial Hygiene of the State of New York publishes the *Industrial Bulletin*, the *Monthly Review*, and special bulletins. The most important contributions have come from L. Greenburg, May R. Mayers, Adelaide Ross Smith, L. J. Goldwater, and H. Heiman, physicians; from Theodore Hatch and A. C. Stern, engineers; and from W. J. Burke, S. Moskowitz, and C. G.

Ford, chemists. Some of the following publications were published in the *Industrial Bulletin,* some in other journals:

Dust

L. Greenburg, A. R. Smith, and J. Seigal. "Respiratory Diseases among Grainworkers," *Industrial Bulletin,* 1941.

L. Greenburg, W. Siegel, and A. R. Smith. "The Dust Hazard in Tremolite Talc Mining" *American Journal of Roentgenology and Radiology,* 1943.

A. C. Stern, J. Baliff, and others. "Characteristics of Unit Dust Collectors," *Monthly Review,* 1946.

Poisons

May R. Mayers and M. McMahon. *Lead in Industry* (Special Bulletin 195, 1938).

May R. Mayers. Carbon Monoxide Poisoning in Industry and its Prevention (Special Bulletin, 1930).

M. R. Mayers and A. R. Smith. "Systemic Effects from Exposure to Certain Chlorinated Hydrocarbons," *Industrial Bulletin,* 1939; *Journal of Industrial Hygiene,* 1939.

C. B. Ford and A. C. Stern. "Occupational Hazards in Fabrication of Magnesium and Its Alloys," *Industrial Bulletin,* 1944.

L. Greenburg, M. R. Mayers, L. J. Goldwater, and A. R. Smith. "Benzol Poisoning in the Rotogravure Printing Industry," *Journal of Industrial Hygiene and Toxicology,* 1939.

H. Heiman and C. B. Ford. "Chronic Benzol Poisoning in the Plastic Industry," *Industrial Bulletin,* 1940.

W. J. Burke and others, "Industrial Air Analysis." (Mimeographed, may be obtained from the Division.)

S. Moskowitz and W. J. Burke. "Control of Health Hazards from Rubber Cementing Operations," *Monthly Review,* 1946.

L. Greenburg and S. Moskowitz. "Occupational Diseases and Hazards in the Chemical and Rubber Industry," *Monthly Review,* 1946.

The Department of Labor of the Commonwealth of Massachusetts publishes yearly reports. In addition, its staff members have written various articles for the *Journal of*

Industrial Hygiene and Toxicology. Among them were studies on chronic benzol exposure by M. Bowditch and H. B. Elkins, 1939; on halogenated hydrocarbons, by L. Levine and H. B. Elkins, 1939; on maximal allowable concentration by M. Bowditch, Alice Hamilton, and others in 1940 and by H. B. Elkins in 1942; and on radium painting, by G. E. Morris, J. R. Tabershaw, and others in 1943 and by J. R. Tabershaw in 1946.

The Department of Labor and Industry of the Commonwealth of Pennsylvania in 1938 published *Survey of Carbon Disulphide and Hydrogen Sulphide Hazards in the Viscose Rayon Industry*, by F. H. Lewey, F. J. Braceland, W. E. Ehrlich, and others. Departments of Labor or Departments of Health in other states also have published valuable research studies.

It may be stressed that in no other country has the government spent so much money and devoted so much effort to scientific research in industrial hygiene as in the United States.

NONGOVERNMENTAL RESEARCH AND STUDIES

Our discussion of scientific literature on industrial hygiene was carried to the middle of the 19th century for Europe and to the second decade of the 20th century for the United States. It was interrupted in order to follow factory legislation, which has developed quickly since that time. In returning to the literature of industrial hygiene and occupational diseases, we shall survey briefly the works produced by scientists not charged with that task by the government.

Physicians get firsthand knowledge of workers' diseases and so are able to discover and describe illnesses caused by new poisons and even illnesses which, though existing for a long time, have been overlooked. Delpech described first the ill effects of carbon bisulphide in the rubber industry

(1856), Grandhomme told of various dangers in the production of tar dyes (1883), Rehn wrote on the bladder carcinoma of aniline workers (1895), and Jaquet dealt with poisoning by methyl bromide (1901). In 1911 John Morley mentioned that S. R. Wilson in a work "at present unpublished" stressed the frequency of scrotal cancer in Manchester.[1] In 1922 Wilson together with A. H. Southam published a report on 69 cases of mule-spinners' cancer seen in the Manchester Royal Infirmary from 1902 to 1922.[2] Since 1919, physicians treating cases of illness caused by mineral oils and its derivatives must report them to the factory inspector. Such cases and the publication mentioned above caused the appointment of a departmental committee, which found that 539 cases of mule-spinners' cancer had occurred since 1876. These are only a few examples of the important role that private physicians have played in this field by making observations in their practice and publishing the facts.

There are other physicians who, without being employed or ordered by any authority, interested themselves in problems of industrial hygiene and collected material laboriously in hospitals or by making study trips and visits to factories.

In 1871 Ludwig Hirt wrote in the introduction to his four-volume work *Die Krankheiten der Arbeiter* (1871–1878), which contains numerous observations that he made himself and is one of the standard works in our field: "I remember only the thousandfold unpleasantnesses and misfortunes, the challenging answers, sometimes containing even more than a 'no,' which a man has to go through, who, quite on his own and without the help of the government and the health authorities undertakes to visit factories,

---

[1] John Morley, "The Lymphatics of the Scrotum," *Lancet*, II (1911), 1545.
[2] A. H. Southam and S. R. Wilson, "Cancer of the Scrotum," *British Medical Journal*, II (1922). 971.

establishments, etc. in order to study the health of workers." May I add my own experience and that of my wife in a later time? In 1906 we looked for phosphorus necrosis in the villages of the Bohemian Forest. After a very exhausting day, during which we had walked for hours to the far-flung houses in which sick people lived, we came late in the evening to the inn of the small village, owned by the match manufacturer. The innkeeper, a tenant, refused to give us a night's lodging, and we had to walk for more than an hour to put up at a very poor inn.

Nevertheless many studies—beginning with those of Ramazzini, Ackermann, and Patissier and later those of Eulenberg, Arlidge, and others—were made in the same or a similar way.

Taking up first those important works which had a great influence on the development of industrial hygiene and which treated the whole field, we find two that were published in England. J. T. Arlidge's *The Hygiene, Diseases and Mortality of Occupations,* London, 1892, a book of 568 pages, is the first comprehensive work on this subject in the English language. The voluminous work edited by Sir Thomas Oliver, *Dangerous Trades,* London, 1902, 981 pages, was written with the collaboration of statisticians, engineers, and physicians, and contains important chapters on lead, phosphorus, earthenware, millstone production, and India-rubber written by the editor himself. Even today it is a very valuable book containing quantities of material.

In Germany we should mention H. Eulenberg's *Handbuch der Gewerbehygiene auf experimenteller Grundlage* (Handbook of Industrial Hygiene on an Experimental Basis) Berlin, 1876, 927 pages, which treats especially of chemical questions and scientific problems. The standard work is, of course, Ludwig Hirt's *Die Krankheiten der Arbeiter; Ein Beitrag zur Förderung der öffentlichen Ge-*

sundheitspflege (The Diseases of Workers; a Contribution to the Advance of Public Health). The four volumes are: I. *Die Staubinhalationskrankheiten und die von ihnen besonders heimgesuchten Gewerbe- und Fabrikbetriebe* (The Dust Diseases and the Trades and Industries Especially Suffering from Them; Breslau, 1871, 304 pages); II. *Die Gasinhalationskrankheiten und die von ihnen besonders heimgesuchten Gewerbe- und Fabriksbetriebe* (The Illnesses Caused by Noxious Gases and the Trades and Industries Especially Suffering from Them; Breslau, 1873, 228 pages); III. *Die gewerblichen Vergiftungen,* etc. (Occupational Poisonings and the Trades; Leipzig, 1875, 289 pages); IV. *Die äusseren* (sc. *chirurgischen*) *Krankheiten der Arbeiter* (The External [that is, Surgical] Illnesses of Workers; Leipzig, 1878, 318 pages). These volumes include clinical pictures and treatment, as well as a great deal on industrial hygiene. It is characteristic of those times that the author could not dedicate his life to this science; he had to leave this unprofitable field and make his living as a practitioner and neurologist.

There is also the section on "Gewerbekrankheiten" (Occupational Diseases), 224 pages, written by Hirt and G. Merkel for the *Handbuch der Hygiene und Gewerbekrankheiten* (Handbook of Hygiene and Occupational Diseases), edited by M. von Pettenkofer and H. von Ziemssen, Berlin, 1882. This work was followed by *Die Gefahren und Krankheiten in der chemischen Industrie und die Mittel zu ihrer Verhütung und Beseitigung* (The Dangers and Diseases in the Chemical Industries and the Methods of Controlling and Removing Them), by C. Heinzerling, Halle, 1886–1887, in two volumes.

In the United States such books were published at a much later date, in accordance with the later development of industries and legislation.

W. H. Tolman, Director of the American Museum of

Safety, in 1913 published in conjunction with L. B. Kendall a book entitled *Safety*, the introduction to which begins: "This book is the only comprehensive work on safety that has yet appeared in the English language." Of its 422 pages, 234 are devoted to accident prevention. The book also treats industrial hygiene and social welfare.

G. M. Price, a physician, Chairman of the Board of the Garment Industry and formerly Director of Investigation of the New York State Factory Commission, is the author of *The Modern Factory* (New York, 1914, 574 pages), a very complete statement of factory hygiene and the legislation controlling it in the United States. As an indication of the frequency of fires in New York and throughout the country, it may be mentioned that 56 pages are devoted to "Factory Fires and Their Prevention," although English handbooks devote only a few pages to this matter, and the most voluminous German books only about 20 pages.

In the same year W. G. Thompson's *The Occupational Diseases* (New York, 1914, 714 pages), was published. In 1916 G. M. Kober and W. C. Hanson, in collaboration with others, wrote *Diseases of Occupation and Vocational Hygiene*, a work of 918 pages. Other books followed, such as the well-known and highly scientific work of Alice Hamilton, *Industrial Poisons in the United States* (New York, 1925, 590 pages).

In discussing the literature on individual poisons and considering first the literature on *lead poisoning*, we must go back to the early years of the 19th century.

In earlier times lead poisoning far surpassed all other occupational diseases in the number of cases and was very widespread. Therefore it is understandable that at the beginning of the 19th century, when attention was given to occupational illnesses, lead poisoning was studied first. In France, F. V. Mérat published *Traité de la colique metallique* (1st ed., 1803; 2nd ed., 1812, 307 pages). Then followed—to mention only the most important publications—the classic work of L. Tanquerel des Planches, *Traité des maladies de plomb ou saturnines* (2 volumes, 1100 pages), published in Paris in 1839 and translated into German by S. Frankenberg in 1842. This work is still very valuable today because of the abundance and exactness of its clinical observations. It deals with 1,213 cases of colic—among them 406 in the manufacture of white lead and 305 cases of painter's colic—101 cases of lead paresis, 72 of encephalopathy, and other cases observed during the years 1831 to 1839 in the Charité. Later J. Renaut published *De l'intoxication saturnine chronique* (Paris, 1875, 198 pages). Madame Dejerine-Klumpke wrote a valuable book, *Des Polyneurites en general et des paralyses et atrophies saturnines* (Paris, 1889, 295 pages). The astonishing concentration in Paris of leadworkers in the most hazardous occupations of their field gave the impetus to these studies on lead poisoning. There were not only many house painters handling lead paints as in every big city, but also factories making lead paints (white lead and minium). Such factories, as a rule, are located outside of large cities.

The next important work on lead poisoning was the well-known book by Thomas M. Legge and K. W. Goadby, *Lead Poisoning and Lead Absorption* (London, 1912, 308 pages), translated into German by H. Katz and L. Teleky with additional remarks concerning German conditions (Berlin, 1921). Legge completed the clinical picture of lead poisoning. Much earlier, Stockhausen (1656) and Tanquerel (1839) had stressed the overwhelming importance of the inhalation of lead dust (as compared with the intake into the stomach from dirty hands). Legge stressed this again, Goadby demonstrated it by animal experiments, and in this way there was built up a truly effective (but more difficult) method of control than the cleanliness recommended by others.

The work of J. C. Aub, L. T. Fairhall, A. S. Minot, and P. Reznikoff, *Lead Poisoning* (Baltimore, 1936, 265 pages), elucidated by modern chemical examinations and studies many problems relating to lead poisoning. Finally we mention the chapter on lead written by F. Flury in the *Handbuch der experimentellen Pharmakologie*, Vol. III, Part 3 (Berlin, 1934). All these books also deal with prophylaxis and control.

So many publications deal with the production and use of white lead and red lead, with typesetting, type foundries, lead and zinc smelters, storage battery manufactures, and other lead-using trades that it is impossible to list here even the most important. Some of these publications were mentioned in the sections on the International Association for Labor Legislation and on governmental research. The student of the subject may refer to two bibliographies. E. Blänsdorf's international bibliography, *Bleiliteratur; Schriften aus dem Gesammtgebiet der Gewerbehygiene* (Vol. VII, Part 2, Berlin, 1922), lists approximately 3,000 publications and 170 regulations. A later work, *Bibliography and Survey of Lead Poisoning*, compiled by F. A. Bronce (New York, 1943), is not nearly so comprehensive as the former.

The literature on mercury poisoning also leads us rather far back—to A. Kussmaul's *Untersuchungen über den constitutionellen Merkurialismus und sein Verhältnis zur constitutionellen Syphilis* (Research on Constitutional Mercurialism and Its Relation to Constitutional Syphilis), Würzburg, 1861, 434 pages. Other books on the subject include the following: Bruno Schönlank, *Die Fürther Quecksilber-Spiegelbelegen und ihre Arbeiter* (The Fuerth Mirror Quicksilvering Factories and their Workers), Stuttgart, 1888, 256 pages; L. Teleky, *Die gewerbliche Quecksilbervergiftung* (Occupational Mercury Poisoning), Berlin, 1912, 225 pages; Mrs. Bates J. W. Linden, *Mercury*

Poisoning in the Industries of New York City and Vicinity, New York, 1912; and E. W. Baader and E. Holstein, *Das Quecksilber und die gewerbliche Quecksilbervergiftung* (Mercury and Occupational Mercury Poisoning), Berlin, 1933, 233 pages. Glibert's book on skins and furs already has been noted.

Books on phosphorus necrosis have been written by Bibra and Geist, *Die Krankheiten der Arbeiter in den Phosphorzündholzfabriken* (The Diseases of the Workers in the Phosphorus Match Factories), Erlangen, 1847, and by L. Teleky, *Die Phosphornekross* (Phosphorus Necrosis), Vienna, 1907, 182 pages.

Books on poisoning caused by carbon monoxide include: W. Sachs, *Die Kohlenoxydvergiftung*, Brunswick, 1900, 236 pages; M. Nicloux, *L'oxyde de carbone et l'intoxication oxycarbonique*, Paris, 1925, 254 pages; L. Lewin, *Die Kohlenoxydvergiftung*, Berlin, 1920, 369 pages; and C. K. Drinker, *Carbon Monoxide Asphyxias*, New York, 1938, 276 pages.

Among books which cover a large part of our field, the most important are the following, arranged chronologically under the various countries. (Some of the titles mentioned above are listed here again in order to give a complete survey.)

INTERNATIONAL

International Labor Office. *Health and Occupation* (French title, *Hygiene du travail*) 1926–1932, 333 chapters, with supplements to 1944.

UNITED STATES

W. G. Thompson. *The Occupational Diseases*. New York, 1914. 714 pp.

G. M. Kober and W. C. Hanson. *Diseases of Occupation and Vocational Hygiene*. Philadelphia, 1916. 918 pp.

G. M. Kober and E. R. Hayhurst. *Industrial Health*. Baltimore, 1924. 1184 pp.

Alice Hamilton. *Industrial Poisons in the United States.* New York, 1925. 590 pages; 2d ed., with R. T. Johnstone, 1916.
Yandell Henderson and H. W. Haggard. *Noxious Gases.* New York, 1st ed., 1927, 220 pp.; 2d ed., 1943.
Alice Hamilton. *Industrial Toxicology.* New York, 1935. 352 pp.
A. J. Lanza and J. A. Goldberg (ed.). *Industrial Hygiene.* New York, 1939. 743 pp.
Rutherford Johnstone. *Occupational Diseases.* New York, 1941. 558 pp.
Alice Hamilton. *Exploring the Dangerous Trades.* New York, 1943. 433 pp. (This autobiography contains chapters on the history of industrial hygiene in the United States.)

Great Britain

J. T. Arlidge. *The Hygiene, Diseases and Mortality of Occupations.* London, 1892. 568 pp.
Thomas Oliver, ed. *Dangerous Trades.* London, 1902. 891 pp.
E. L. Collis and Major Greenwood. *The Health of the Industrial Worker.* London, 1921. 450 pp.
E. W. Hope, W. Hanna, and C. O. Stallybrass. *Industrial Hygiene and Medicine.* London, 1923. 766 pp.
T. M. Legge. *Industrial Maladies,* ed. by S. A. Henry. London, 1934. 234 pp.
Ethel Browning. *Toxicity of Industrial Organic Solvents.* London, 1937. 366 pp.

Germany

L. Hirt. *Die Krankheiten der Arbeiter.* 4 vols., Breslau and Leipzig, 1871–1878.
H. Eulenberg. *Handbuch der Gewerbehygiene auf experimenteller Grundlage.* Breslau, 1876. 927 pp.
C. Heinzerling. *Die Gefahren und Krankheiten der chemischen Industrie.* 2 vols., Halle, 1886–1887.
Theodor Weyl, ed. *Handbuch der Hygiene,* Vol. VIII: Gewerbehygiene. Jena, 1st ed., 1897, 1255 pp.; 2d ed., 1913.
Theodor Sommerfeld. *Handbuch der Gewerbekrankheiten.* Berlin, 1898. 536 pp.
K. B. Lehmann. *Kurzes Lehrbuch der Arbeits- und Gewerbehygiene.* Leipzig, 1919. 468 pp.
A. Gottstein, A. Schlossmann and L. Teleky, ed. *Handbuch der*

Sozialen Hygiene und Gesundheitsfürsorge. Vol. II: Gewerbekrankheiten und Gewerbehygiene, Berlin, 1926. 816 pp.

F. Flury and F. Zernik. *Schädliche Gase.* Berlin, 1931. 637 pp.

Franz Koelsch. *Handbuch der Berufskrankheiten.* Jena, 1935 and 1937. 1175 pp.

Frank Koelsch. *Lehrbuch der Arbeitshygiene.* Stuttgart, 1946–47. 2 vols. 367, 483 pp.

FRANCE

A. Layet. *Hygiène des professions et des industries.* Paris, 1875. 560 pp.

Office du Travail. *Poisons industriels.* Paris, 1901. 449 pp.

Georges G. Paraf. *Hygiène et sécurité du travail industriel.* Paris, 1905. 632 pp.

ITALY

G. Pieraccini. *Patologia del lavoro e terapia sociale.* Milan, 1906. 695 pp.

A. Ranelleti. *Le Malattie del lavoro.* Rome, 1924. 320 pp.

N. Castellino. *Il Lavoro nella chimica industriale.* Milan, 1940. 390 pp.

L. Preti, ed. Trattato di patologia medica del lavoro. Milan, 1940. 832 pp.

NETHERLANDS

L. Heijermans. *Handleiding tot de Kennis der Beroepsziekten.* Rotterdam, 1908. 550 pp.

This enumeration not only lists the most important books but also indicates the leading scientists in the field who have carried out a great deal of research. The list would be incomplete without mention of the Swiss Heinrich Zangger, for decades the leading scientist in industrial hygiene of his country. Besides works on toxicology he wrote numerous articles on industrial hygiene (carbon monoxide, arseniureted hydrogen, electric accidents, celluloid).

Journals. Scientific articles on industrial hygiene and

occupational diseases have been published in all of the better medical or technical journals, such as the *Journal of Hygiene* (England), the *Archiv für Hygiene* and the *Zeitschrift für Hygiene und Infectionskrankheiten* (Germany), and the *Annales d'hygiène publique* (France). In this century a number of journals have been devoted exclusively to our field, publishing numerous and valuable articles. The following is a list of these journals and the years in which they appeared. How many of them survived World War II, other than those published in England and the United States, is difficult to say—also which will be revived. The first such journals appeared in Austria. Then followed some important ones in Italy.

Austria

Zeitschrift für Gewerbehygiene, Unfallverhütung und Arbeiter wohlfahrts Einrichtungen. Vienna, 1893– .

Italy

Ramazzini. Florence, 1907–1914.
Il Lavoro. Milan, 1908–1924; since 1925, *Medicina del lavoro.*
Rassegna di medicina applicata al lavoro industriale. Turin, 1930–1938; since 1939, *Rassegna di medicina industriale.*

Germany

Mitteilungen des Institutes für Gewerbehygiene. Frankfort on the Main, 1910–1912.
Zentralblatt für Gewerbehygiene und Unfallverhütung. Berlin, 1913–1943.
Beihefte zum Zentralblatt für Gewerbehygiene und Unfallverhütung. Berlin, 1913–1939.
Schriften aus dem Gesamtgebiet der Gewerbehygiene. Berlin, 1913–1935.
Arbeitsschutz, Unfallverhütung, Gewerbehygiene (Sonderausgabe des Reichsarbeitsblattes, Part III). Berlin, 1925– .
Arbeit und Gesundheit (Schriftenreihe des Reichsarbeitsblattes). Berlin, 1925–1943.

Archiv für Gewerbepathologie und Gewerbehygiene. Berlin, 1930–1944.
Arbeitsmedizin. Leipzig, since 1935; No. 23 in press.

UNITED STATES

Journal of Industrial Hygiene. Baltimore, 1919–1935; since 1936, *Journal of Industrial Hygiene and Toxicology.*
Industrial Medicine. Chicago, 1930– .
Occupational Medicine. Chicago, 1946– .
American Industrial Hygiene Association Quarterly. Chicago, 1946– .

FRANCE

La Médicine du travail. Lyon, 1929– .
Archives des maladies professionelles, de médicine du travail et de sécurité sociale. Paris, 1938– .

ENGLAND

British Journal of Industrial Medicine. London, 1945– .

Congresses. In 1905 an Italian committee headed by Luigi Devoto, the chief of the clinic for occupational diseases in Milan, called the First International Congress for Occupational Diseases, which met in Milan. Since that time, other such congresses have been called together by the International Permanent Commission for the Study of Occupational Diseases. These congresses became the meeting place of all specialists in this field. The same is true of the congresses convoked by the Permanent International Committee for the Study of the Medicine of Accidents, after it had founded a section for occupational diseases. The congresses were: Milan, 1905; Brussels, 1910; Vienna, 1914 (not held because of World War I, but the lectures prepared for it have been published); Amsterdam, 1925; Budapest, 1928; Lyon, 1929; Geneva, 1931; Brussels, 1935; and Frankfort on the Main, 1938.[3] The protocols of all these congresses contain many valuable articles.

[3] The congresses held at Brussels (1910) and Vienna were convoked by the Permanent Commission for the Study of Occupational Diseases; that held at

In addition to these international congresses, there were national ones, such as the Congress de la Hygiène des Travailleurs et des Ateliers, held annually since 1904 by the Confédération Général du Travail in France. I do not know if they survived World War I. Beginning in 1907, a committee of Italian physicians biennially called a Congresso Nazionale per le Mallatie del Lavoro. These congresses continued into World War II.

Mention also should be made of the International Congresses for Rescue Work and First Aid, one of them called "for Rescue Work and Accident Prevention," although they scarcely treated prevention of industrial accidents. The first took place in Frankfort on the Main in 1906, the second in Vienna in 1913, the third in Amsterdam in 1926, and the fourth in Copenhagen in 1934.

Exhibitions and Museums. The development of practical industrial hygiene also has been promoted by exhibitions. The International Exhibition for Health and Rescue Work in Brussels (1876) seems to have been the first in this field. Others followed: one in Berlin (1883), a section of the World's Fair in Brussels (1910), the Internationale Hygiene Ausstellung in Dresden (1911), the Ausstellung für Gesundheitspflege, soziale Fursorge, und Leibesubungen ("Gesolei") in Dusseldorf (1928), and a section of the World's Fair in New York (1939).

Even more important than these in spreading knowledge and in calling attention to new constructions and inventions were the permanent exhibits:

Gewerbehygienisches Museum, Vienna, 1890–1909, thereafter part of the Technical Museum.
Bayerisches Arbeiter Museum, Munich, founded in 1900.
Ständige Ausstellung für Arbeiterwohlfahrt, Berlin-Charlottenburg, founded in 1903.

Budapest was convoked by the Permanent International Committee for the Study of the Medicine of Accidents; and these two organizations jointly convoked the congresses held at Geneva, Brussels (1935), and Frankfort. The remaining congresses were called by special committees.

American Museum of Safety, New York, founded in 1911 and closed in 1920.

Industrial Hygiene Museum, a section of the Deutsches Hygiene Museum, Dresden, founded as a permanent continuation of the Hygiene Exhibition in 1912.

Home Office, Industrial Museum in London, authorized by Parliament in 1913 and opened in 1927.

9

The Progress in Other Sciences and Its Influence on Industrial Hygiene

WE HAVE DISCUSSED the legislation for industrial hygiene and the circumstances as well as the powers by which it was instigated and by which research work in industrial hygiene was stimulated. It would not have been possible to achieve a great deal of this or even to recognize clearly the existing evils if there had not been immense progress in other sciences and techniques. We shall discuss this progress in so far as it influenced industrial hygiene.

TOXICOLOGY

Toxicology has advanced greatly in the last decades, settling many questions by the use of newer methods and the performing of definite animal experiments as well as chemical tests. The first toxicologic examinations for the purposes of industrial hygiene to be made on a large scale were those of K. B. Lehmann. Beginning in 1886, the experiments were continued by him and by his pupils for several decades. Results were published in the *Archiv für Hygiene* under the title "Experimentelle Studien über den Einfluss technisch wichtiger Gase und Dämpfe auf den Organismus" (Experimental Studies Concerning the Influence of Technically and Hygienically Important Gases and Vapors on Organism).

Some of the fundamental works on toxicology are:

GERMANY

L. Lewin. Lehrbuch der Toxikologie. 1st ed., Leipzig, 1885, 456 pp. 4th ed. (called *Gifte und Vergiftungen*), Berlin, 1929, 1019 pp.

R. Kobert. Lehrbuch der Intoxicationen. 1st ed., Stuttgart, 1893, 816 pp. 2d ed., 2 vols., 1902–1906: Vol. I, 302 pp., Vol. II, 1298 pp.

A. J. Kunkel. Handbuch der Toxikologie. Jena, 1901. 1117 pp.

F. Erben. Vergiftungen. In two parts (Vol. VII of *Handbuch der ärztlichen Sachverstandigen-Tätigkeit*): Part I, Vienna, 1909, 458 pp.; Part II, Vienna, 1910, 1245 pp.

H. Zangger. Vergiftungen. In *Diagnostische und therapeutische Irrtümer und deren Verhütung*, ed. J. Schwalbe. Leipzig, 1924.

F. Flury and H. Zangger. Lehrbuch der Toxikologie. Berlin, 1928. 500 pp.

E. Starkenstein, E. Rost, and J. Pohl. Toxikologie. Vienna, 1929.

E. Petri. Pathologische Anatomie und Histologie der Vergiftungen. Vol. X of *Handbuch der speciellen Pathologischen Anatomie und Histologie*. Berlin, 1930. 640 pp.

England

A. W. Blyth and M. W. Blyth. *Poisons, Their Effects and Detection*. 1st ed., 1884; 5th ed., 1920.

France

P. M. Orfila. *Traité des poisons*. 4 vols., Paris, 1814–1815.

A. Tardieu. *Etude medico-legale et clinique sur l'empoisonnement*. Paris, 1st ed. 1867, 1072 pp.; 2d ed., 1875, 1235 pp.

P. Brouardel. *Les Empoisonnements criminels et accidentels*. Paris, 1902.

P. Brouardel. *Les Intoxications*. Paris, 1904.

P. Carnot and others. *Intoxications*. Paris, 1907.

United States

C. M. Riley. *Toxicology*. St. Louis, 1902. 132 pp.

J. W. Holland. *A Textbook of Medical Chemistry and Toxicology*. Philadelphia, 1906. 592 pp.

A. H. Brundage. *Toxicology*. 1st ed., Brooklyn, 1911; 15th ed., New York and London, 1926.

T. H. Sollmann. *A Manual of Pharmacology and Its Application to Therapeutics and Toxicology*. Philadelphia, 1st ed., 1917, 901 pp.; 6th ed., 1942, 1298 pp.

W. D. McNally. *Toxicology*. Chicago, 1937. 1022 pp.

HYGIENE AND PHYSIOLOGY

It is self-evident that the progress of hygienic science as a whole should also promote industrial hygiene. The research done on the effects of the atmospheric contents on man has been especially valuable. The work of Leonard Hill, *The Science of Ventilation and Open Air Treatment*, London, 1919, and his investigations on the influence of temperature, humidity, and air movement, as well as the construction of the katathermometer, made it possible to recognize the atmospheric conditions and to evaluate them in their influence on the worker. The studies made by the American Society of Heating and Ventilating Engineers together with the United States Bureau of Mines were also extremely helpful. R. R. Sayers, C. P. Yaglou, and F. C. Houghten determined the "effective temperature" and the "comfort zone" by summarizing all the factors of the surrounding air to which men react.

Technical improvements enabled engineers to put to a practical test what had been established theoretically and thus to comply with hygienic requirements. The development of ventilation and heating cannot be discussed here, but the extent of the practical achievements in this field may be recognized readily when we consider the "air-conditioned" rooms which exist in some offices and even in some factories.

The progress made in the hygiene of vision and illumination and also in transportation (elevators, band conveyors) is no less remarkable. The technical achievements in the construction of closed disintegrators and closed means of transporting dusty or evaporating material are of the greatest value for industrial hygiene.

STATISTICS

Much research in industrial hygiene and many investigations into the dangers of certain trades are based on statistics

THE PROGRESS IN OTHER SCIENCES 183

on mortality and morbidity as a whole or on those of individual illnesses. Therefore we must examine the history and the development of statistics concerning mortality and morbidity in occupations. As it is not possible to treat fully all of the publications dealing with this problem, we shall mention only the most important ones and at the same time point to the progress that has been made in statistical methods.

The earliest published work on vital statistics is Johann Peter Süssmilch's *Die göttliche Ordnung in den Veränderungen des menschlichen Geschlechts* (The Divine Order in the Changes of Mankind), 2 volumes, Berlin, 1761–1762, 1391 pages. This study reveals a higher annual mortality rate in large cities than in rural areas: in Rome, one in every 24 or 25 of the population; in Berlin, one in 28; in small towns, one in 32; but in rural regions, one in 40. Discussing the reasons for this difference, he points to the immorality in cities, to overeating and excessive drinking, the crowding of many persons into insufficient quarters, the fumes of the many fireplaces, and to poor housing and epidemics—but he does not mention the influence of the trades. He deals very extensively with the advantages and disadvantages of industry for the state. "Industry" is for him identical with the manufacture of textiles, mainly cotton. He states that "the peasant is stronger and better able to resist wind, weather and hardships than those who work in warm rooms at the loom"—but he does not give the mortality rate among the latter.

In 1803 Georg Adelmann published a book on diseases of artists and artisans, based on statistical tables set up from 1786 to 1802 by the institute for sick journeymen, artists, and artisans in Würzburg, Germany.

At the beginning of the 19th century, scientists tried to learn more about occupational mortality. However, the only data available concerned those who had died; there were no figures as to the number in the same occupation who were still living. Several authors tried to obtain more exact statistics, but others contented themselves with tables that showed the average age at death, sometimes incorrectly labeling the results as the "probable duration of life." Casper in his book, *Die wahrscheinliche Lebensdauer* (The Probable Duration of Life), Berlin, 1835, compiled such tables for ministers, higher and lesser officials, farmers, artists, and teachers. The following books dealing with this matter are also worthy of note: Lombard, *De l'influence des professions sur la durée de la vie* (The Influence of Occupations on the Duration of Life), Geneva, 1835; W. C. de Neufville, *Lebensdauer und Todesursachen Zwei-und-Zwanzig verschiedener Stände und Gewerbe* (The Duration of Life and the Causes of Death in Twenty-two Different Professions and Trades), Frankfort on the Main, 1855; Escherich, *Hygienisch-statistische Untersuchungen über Lebensdauer in verschiedenen Ständen (Geistliche, Lehrer, Beamte) in Bayern* (Hygienic-statistic Investigations Concerning the Duration of Life in Different Professions [Ministers, Teachers, Officials] in Bavaria), Würzburg, 1854.

Despite the incompleteness of the material and the incorrect use made of it, and although the figures are not at all exact, the authors were able to make valuable deductions as to the differences between the professions. Casper found the average age of theologians at death to be 65.1 years and that of physicians 56.8. Escherich came to similar conclusions. For groups of a hundred, each, in several occupations, Neufville found that among the ministers who had died, only two were between 20 and 29 years of age, whereas the percentage of shoemakers who died at this age was 30.9, and of lithographers, 43.5. Among the tailors, 42.8 percent had died of tuberculosis, but only 8.2 percent of the butchers and 6.8 percent of the jurists had died from this disease.

Finally there are several other extensive works: F. Oesterlen's two-volume *Handbuch der medizinischen Statistik* (Handbook of Medicinal Statistics), Tübingen, 1867, with a second edition in 1874; and Oldendorff's *Der Einfluss der Beschäftigung auf die Lebensdauer der Menschen* (The Influence of Occupation on the Duration of Life of Men), Berlin, 1878, in two volumes.

Several publications of the next decades deal with individual trades in a more exact way. These include J. Sendtner's booklet on the mortality of beer-brewers, Munich, 1891; a booklet by A. Geissler on the mortality of physicians of Saxony, published in Leipzig in 1887; and an article by W. Weinberg on the mortality of physicians of Württemberg, published in the *Württembergische Jahrbücher für Statistik und Landeskunde, 1897*. In addition, there are studies based on the material of the Gothaer Lebensversicherungsbank (Gotha Life Insurance Bank); Karup, Gollmer, and Florschütz, *Aus der Praxis der Gothaer Lebensversicherungsbank* (From the Practice of the Gotha Life Insurance Bank), Jena, 1902, and other, later, material from the same source. A Andrae's articles on the mortality of persons engaged in the production and sale of alcoholic beverages (1905) and in agricultural occupations (1906), both published in the *Zeitschrift für die gesammte Versicherungswissenschaft*. Similar statistics were published by the life insurance companies of other countries.

In Great Britain, the records of the Friendly Societies provided material on the mortality of occupations. The topic was first treated by F. G. P. Neison in an article published in the *Journal of the Statistical Society*, in 1845, and later in his book, *Conditions of Vital Statistics*, London, 3d ed., 1857. In the statistical periodicals of that time, especially in the English *Journal of the Statistical Society*, the mortality in trades and professions was one of the principal subjects. Among the contributors to the *Journal* was W. A. Guy, who wrote, in addition to other articles, "On the Duration of Life in the Members of the Several Professions" (1846). It may be recalled that in those days statistics were highly favored in Europe, and their findings, not infrequently based on unreliable research, were overestimated—just as they are today in other countries.

We will not enumerate here all the statistical publications on occupational mortality. In Europe these are now based largely on government figures (to be discussed later) and therefore avoid the mistakes caused by the inadequacy of the data on which the earlier studies were based. We refer instead to the comprehensive books of Westergaard, *Mortalität und Morbidität*, Jena, 1882, 504 pages; 2d ed., 1901, 703 pages; Prinzing, *Handbuch der Medizinischen Statistik*, 1906, 559 pages; 2d ed., 1930–1937, 672 pages; L. Teleky, *Vorlesungen über Soziale Medizin* (Lectures on Social Medicine), Jena, 1914, 282 pages; and Collis and Greenwood, *The Health of the Industrial Worker*, London, 1921, 450 pages.

In the United States, until the last few years, statistics have been based on the records of the big life insurance companies. There are two publications which deal with deaths only, without indicating the number of living persons in the groups in which the deaths occurred, because those figures were not available. These are L. I. Dublin's study in 1917 (Bulletin No. 207 of the Bureau of Labor Statistics) and a work in 1930 by Dublin and R. J. Vane, "Causes of Death by Occupation" (Bulletin No. 507). Both studies are based on the mortality experiences of industrial policyholders insured with the Metropolitan Life Insurance Company, the first for the years 1911–1913, the latter for 1922–1924. Figures given in these bulletins show that there were 94,269 and 105,467 deaths respectively among white males. Without giving the figures on the corresponding living persons, the authors succeeded in making as much as possible from this selected material by using suitable methods and by taking into consideration statements from the literature. Other publications based on the material of life insurance companies were published by the Actuarial Society of America and the Association of Life Insurance Medical Directors. They are *Joint Occupation Study 1928,* which took into account 1,300,000 insured persons and 22,600 fatalities (1915–1926), and the *Occupational Study* of 1937, which completes the first.

Two other publications, written by E. E. Cammack, chairman of the American committee on group mortality investigation, cover the years 1913 to 1926 and also 1921 to 1926 and 1927 to 1931. The second publication deals with 14,000,000 life years of insured persons and 115,621 "cases of death including cases of permanent total disability before the age of 60 years." To treat as a whole the cases of death and total disability (the latter comprising 10 to 20 percent of all the cases) may be justified for purposes of insurance calculations but diminishes the scientific value of the results.

The official statistics of several countries include tables on occupational mortality, based on the figures for living persons according to the census and figures of deaths from the official reports. But there are some difficulties involved. The *Statistisches Jahrbuch für den Preussischen Staat* (Statistical Yearbook for Prussia) for 1911 gives a table of statistics on occupational mortality in Prussia for the years 1906 to 1908, but the trade groups are very large (there are only 19 in all), and no distinction is made between employers and employees. The census of Switzerland classed a worker with the industry in which he was employed; for instance, a repair locksmith working in a textile factory was classed as a textile worker. But in the death registers he was listed as a locksmith. Thus the picture of occupational mortality is inaccurate.

The Netherlands published similar statistics for the years 1891 to 1895 and 1908 to 1911, and planned to do this every ten years. The totals in some occupations are very small, although the deaths of a four-year period are lumped together.

The first and until now the only American work based on the census and the official death register, both of the year 1930, is that of Jessamine S. Whitney, statistician for the National Tuberculosis Association, *Death Rates by Occupation*, New York, 1934. The death registers of ten states —states in which 39 percent of the gainfully occupied males in the United States are employed—are used. The registrars in these states were given specific instructions. There are tables on the mortality of social groups and also on the mortality in 54 occupations with more than 500 deaths each. Figures are given for three age groups and ten groups of diseases. This is a fine work, and we hope that many more will follow.

The oldest and by far the best government statistics on occupational mortality are the British. It was W. Farr who, in the fourteenth annual report of the Registrar General (1857), made the pioneer contribution to the subject.[1] In 1864 he published statistics on occupational mortality, using the population records of the 1861 census and the death registers from the years 1860 and 1861. Every ten years subsequently, occupational mortality has been computed very carefully, using the census reports and the death records for those years preceding and following the census, and the results have been published as a supplement to the annual report of the Registrar General. In the first report

[1] M. Greenwood, "Occupational and Economic Factors of Mortality," *British Medical Journal*, I (1939), 862.

and in that of 1871, Farr dealt with mortality as a whole and at certain ages in connection with certain occupations. Ogle (1880–1882) took into account specific causes of death. The reports were improved and enlarged in every decennial edition. Occupations of male persons are now divided into a great number of categories, those of women into a smaller number, and 47 causes of death are listed which for certain purposes are condensed into 24 groups. Persons are divided into nine age groups. All the important points and results are discussed thoroughly. These supplements are thus the fundamental source of our information on occupational mortality.

The beginning of statistics concerning the *morbidity* of occupations goes very far back. In Scotland, according to Westergaard the Highland Society promised prizes in 1820 for good reports from sick funds (Friendly Societies). The first publications based on these reports were written by Oliphant (1824), then by the above-mentioned Neison. A. G. Finlaison published *Return on Sickness and Mortality Experience of Friendly Societies* (London, 1853 and 1854). A. W. Watson investigated the sickness and mortality reports of I.O.O.F., Manchester Unity, for the years 1893 to 1897 (Manchester, 1903).[2]

The laws that introduced obligatory health insurance in Germany (1883) and Austria (1888) imposed upon sickness funds the obligation to supply statistical reports to the government. Particularly the Austrian law and rules asked for comprehensive statistics. But the reports were so unsatisfactory that the government published their findings only once. The Verband der wiener Genossenschaftskrankenkassen (Union of the Viennese Guild Sick Funds) and the Allgemeine Arbeiter Kranken- und Unterstützungskasse (General Workers' Sick and Benefit Fund), the

[2] E. L. Collis and M. Greenwood, *The Health of the Industrial Worker* (London, 1921), p. 66–69.

first comprising the smaller workshops, the latter the factories, issued detailed statistics up to the year 1904. These were thoroughly elaborated by S. Rosenfeld, who published his work in the *Statistische Monatsschrift* in 1905 and the following years.

In Germany, official reports contain nothing on the morbidity of occupations. In 1900 the *Frankfurter Krankheits-tafeln* (Morbidity Tables of Frankfort) were published. In 1910 a great work was published by the Kaiserliche Statistische Amt: *Krankheits- und Sterblichkeitsverhältnisse in der Ortskrankenkasse für Leipzig und Umgebung* (Morbidity and Mortality in the Sickness Fund for Leipzig and Surroundings), which treated the material collected during the years 1887 to 1905. Many sickness funds published statistics, such as *Krankheitsstatistik der Allgemeinen Ortskrankenkasse der Stadt Berlin* (Morbidity Statistics of the General Local Sickness Fund of Berlin), 1915–1918, worked out in the German Health Office. In the year 1925, B. Heymann and F. Freudenberg published *Morbidität und Mortalität der Bergleute im Ruhrgebiet* (Morbidity and Mortality of the Miners in the Ruhr district), Essen, 200 pages, based on the statistics of the Allgemeinen Knappschaftsverein in Bochum (General Miners' Association in Bochum).

In 1921, the sick funds of the Rhineland agreed on a common scheme for reporting illnesses and another for occupations, on which uniform statistics should be based. Their results were published by L. Teleky for the year 1923 in the *Reichsarbeitsblatt*, 1924; for the five-year period from 1922 to 1926 in the *Reichsarbeitsblatt* supplement for 1929; and for the years 1922 to 1931 in the *Archiv für Gewerbepathologie*, Volume V (1934). The old statistics and even the famous one of the Leipziger Ortskrankenkasse show the mistakes which arose when the peculiarities of the sickness funds returns were not taken into consideration. The sick funds of the Rhineland tried to avoid this pitfall in their statistics and publications. To eliminate careless reports and returns grudgingly submitted, the submission of these statistics was made voluntary with the individual sickness funds. Diseases were grouped in such a way that several single diagnoses which usually were made without sharp differentiation had to be put into the same group. For example, one group was "tuberculosis and suspicion of tuberculosis," into which were put also all kinds of apicitis. Another group was "illnesses of the stomach, intestines, and liver"; another, "rheumatism of muscles, of joints and gout." An alphabetic catalogue indicated in which group each of the various diagnoses should be placed. A similar catalogue was made up for the different trades and occupations. Because of the different distribution of the ages in various trades, standardization was necessary. In the publication of these statistics it was stressed that sick fund reports and statistics are not an infallible gauge of the number of cases of disabling illness, as is supposed. The figures necessarily include only those cases of disabling illness actually reported, among which are some cases of pretended illness. Apart from a few serious or painful cases, whether a worker reports himself as disabled depends to a great extent on accompanying circumstances: the exact proportion between the sick benefit and his salary, the method of checking reports of sickness, and other special economic factors. All these factors are different in the different sick funds and at different times. The publications do not summarize the reports of a number of different funds; instead, they point out differences in health figures of various trades within a single fund. If similar results are found for the same trade in several different funds, only then are conclusions drawn on the health of that trade. For example, among all trades, printing held a very favorable position, in all the funds with the textile trades a close second. Metalworking had an unfavorable health rating.

The statisticians of the United States Public Health Service, W. M. Gafafer, D. K. Brundage, and R. H. Britton, published many reports on sickness among members of factory sick funds, in which they tried to make the best use of the statistics collected. To a certain extent their reports are no doubt a reliable basis for recognizing changes in morbidity, even of individual illnesses, according to seasons and calendar years. But the material does not give reliable data for occupational morbidity. There are—in addition to the circumstances mentioned above—too many artificial conditions. Most of the sick funds grant benefits only for illnesses entailing more than six days' disability. The importance of this restriction may be recognized in an article by D. K. Brundage which shows that among all absences of persons employed for less than five years, 67.5 percent lasted from one to three days; 20.2 percent lasted four to seven days; 12.3 percent lasted eight days and longer.[3] Furthermore, benefits are refused for disability caused by narcotics, "immoral practices," venereal diseases, and unlawful acts. Some sickness funds exclude a long list of illnesses: neuritis, lumbago, lame back, rheumatism, and pre-existing diseases. Some plants make insurance in the sick fund compulsory for all workers of the plant; some have pre-examinations, others do not.[4] All of this makes any comparison of statistical data on the subject in this country even more difficult than in other countries. Authors indeed rarely try to make such comparisons, and then in a restricted way. It is evident that the use of statistics of

[3] Dean K. Brundage, "Sickness among Persons in Different Occupations of a Public Utility," *Public Health Reports*, XLIII (1928), 318; Reprint No. 1207.
[4] *Ibid.*, "A Survey of the Work of Employees' Mutual Benefit Associations," *Public Health Reports*, XLVI (1931), 2102–119; Reprint No. 1506.

the American sick funds involves especially great difficulties, more than those of other countries in which the funds are more or less standardized by law.

It is to be hoped that future health insurance in Britain, as provided in the National Health Service Act, will furnish reliable statistics in consequence of the size and uniformity of the organization. Perhaps this may be true also of the planned compulsory health insurance in the United States.

The development of mortality statistics on occupation shows that the most serious problem is to procure the figures on the corresponding living persons and to see that the enumeration of both dead and living is handled in an identical way. If this is accomplished, as it was in the English statistics, the main source of error lies in the unreliability of some diagnoses made by the physician or, in some countries, by the coroner.

It is much more difficult to develop reliable morbidity statistics of occupations. This can be done only on the basis of sickness funds reports.

There are numerous obstacles to the interpretation of these reports, even in countries with compulsory health insurance. The most important difficulty is caused by the fact that we cannot deal with cases of disabling illness but only with *reports* of disability. Reporting depends on many accessory circumstances.

Slowly we have come to recognize that figures alone do not give reliable answers, either in the mortality or in the morbidity statistics of occupation. Research will always be necessary to explain the background of the figures, that is, to ascertain how far they are influenced by occupation, by social conditions, or by selection. Only when these questions are settled will a solid basis be laid for the activity of the industrial hygienist.

10

Special Problems

DUST DISEASES

THE HISTORY of miner's phthisis and that of stonecutters goes far back. Agricola wrote in his book *De re metallica* (1556): "Other mines are dry and the constant dust enters the blood and lungs, producing difficulties in breathing. When dust is corrosive it produces consumption. In Altenberg, Meissen, the men bind loose coverings (bladders) to their faces, so the dust is not carried into the lungs and blood and does not hurt the eyes." Paracelsus too (1567) described the *Bergsucht* (miner's phthisis) and traced it back to astral influences, evaporations of the rocks, and rotten air. Ursinus (1652) considered dust inhalation the cause of "peripneumonia." Stockhausen (1656) also spoke of the *Bergsucht*. Diemerbroeck in his *Anatome corporis humani* (1683) described the autoptic findings in a stonecutter's lung. Löhneiss (1690) said with reference to miners: "The dust and stones fall upon the lung, the men get lung diseases and at last take consumption."[1] A patent for grinding flints by a wet method, granted in 1713 to Thomas Benson of New Castle under Lyme, stated that the dry process proved very destructive to mankind in that no person, however healthy and strong, working in that industry, could possibly survive more than two years because of the dust sucked into his body in the air he breathed.[2] Ramazzini (1700) differentiated the various kinds of dust, and pointed out that not all of them cause consumption. J. F. Henckel

[1] G. E. Löhneiss, *Bericht von Bergwerken* (Stockholm and Hamburg, 1690), p. 56.
[2] E. L. Collis, *Industrial Pneumoconioses; Milroy Lectures, 1915* (London, 1917), p. 3.

(1728) discussed "peripneumonia montana," which eventually may be complicated by phthisis fever. As the external causes of the illness he names lack of air, chokedamp, and bad posture during work. The "stonecutter disease" was dealt with by the Germans Joannes Bubbe (1724) and Wepfer (1728), the famous Swedish scholar Linné (1734), and the Frenchmen Boisin de Sauvages (1772) and Leblanc (1775). Ackermann (1781) discussed the illness extensively, stating that among the tradesmen those who suffer most from the dust are those who must inhale a sharp-edged, insoluble dust such as the dust of stones. Thus even at the end of the 18th century, scientists realized the noxiousness of dust, the differences in the effect of various dusts, and the development of "dust lung" into phthisis.

In the first decades of the 19th century, publications dealing with stonecutters were written by several German authors: W. L. Götzinger (1804 and 1812), Schumann (1821), and Petrenz (1844). The Englishman Thackrah wrote in 1832: "In the mines of the North of England the men are injured by working [lead] ore in sandstone, but are sensible of no inconvenience when the ore is in limestone." [3] Calvert (Holland, 1843) discussed the terrible phthisis mortality rate among fork grinders who worked on dry sandstone.

In the British Royal Commission, appointed in 1862 to inquire into the health of men employed in metalliferous mines, the statistician Farr and the physician Peacock stressed the high mortality from respiratory diseases. Peacock outlined the differences between common and miner's phthisis. Medical men working in the mining district emphasized that the miners' disease differed from ordinary consumption or tuberculous phthisis: the former occurred later in life, it was associated with breathlessness and asth-

[3] C. Turner Thackrah, *The Effects of Arts, Trades and Professions on Health and Longevity* (2d ed.; London, 1832), p. 89.

matic symptoms, and it did not show the same marked incidence among the families of those affected. But, notwithstanding the evidence of several witnesses, the report of the Commission holds that the principal causes of miner's phthisis are exposure to the fumes of explosives and fluctuations of temperature. Dust was considered only subsidiary to many other conditions. E. L. Collis, who mentioned these views in his Milroy lectures, thinks that the commission came to this conclusion because at that time many investigations were being made into bad housing, lack of ventilation, and poverty as the main causes of phthisis.

Before the beginning of the 18th century, coal mines were of importance only in England. The mines on the Continent produced zinc ore, lead ore, silver ore, and gold. But in England, by 1660, 2,140,000 tons of coal already had been extracted. By 1800 the entire Ruhr district had produced 230,558 tons of coal, and by 1850 2,600,000 tons of coal had been extracted in all the Prussian coal mines. With the growing importance of coal mining, the changes in the lungs of coal miners aroused the interest of physicians. George Pearson (Great Britain) was the first to publish his findings on the lungs of coal miners (1813), and was followed by Gregory (1831). The two Thomsons, the elder and the younger, made autopsies of such cases (1824 and 1825) and published their results in 1837. Other authors followed. The word "anthracosis" was first used by Stratton (1837). In Germany, Erdmann of Dresden described the changes in the lungs of coal miners (1831).[4] Brockmann (1844) described in his book *Die metallurgischen Krankheiten des Oberharzes* (The Metallurgical Diseases of the Upper Harz), 1851, similar changes in the lungs of ore miners, a melanosis caused by the fumes of explosives and mine lamps. He reported on pigment corpuscles with a

[4] For an extensive history of the knowledge of dust lung, especially in Germany, see L. Teleky, "Studien über Staublunge," *Archiv für Gewerbepathologie*, III (1932), 418–70.

diameter of 2.4 to 3 microns in the lungs. Chemical examination also showed silica in the ashes of the lungs.

While the English physicians always reported the finding of particles of coal in the lungs and saw in them the cause of anthracosis, there was no unanimity in this respect on the Continent. The Frenchmen Béhier (1837), Rillet (1838), Tardieu (1850), and Bouillard (1861) were of the same opinion as the English. On the other hand, the Frenchmen Brechet (1821), Andral and Grisolle, Trousseau and Leblanc (1828), as well as the Germans Heusinger, Becker, Hesse, and above all, the leading pathologist, R. Virchow (1847), and the leading anatomist, Henle (1862), and Friedrich (1864) held that the dark masses in the lungs were a pigment coming from the blood under abnormal circumstances. This opinion was shaken in 1860 when Traube found splinters of charcoal in the melanotic lungs of a man and again in 1867 when Zenker found lungs red with iron oxide. These findings changed even Virchow's mind, and he admitted the presence of coal in anthracotic lungs.

In England, H. H. Greenhow published observations and pathologic findings on many dust workers: miners, needle pointers, cotton operatives, flax hacklers, and metal grinders (1862–1870). In Germany, F. Meinel published *Ueber die Erkrankungen der Lungen durch Kieselstaubinhalation* (On the Diseases of the Lungs by Inhalation of Silica Dust), 1869, and Seltmann wrote *Die Anthrakosis der Lungen bei den Kohlenbergarbeitern* (Anthracosis of the Lungs in Coal Miners), 1867. The latter comes to the conclusion that high degrees of coal impregnation can occur without serious symptoms, while stone dust containing quartz is highly injurious. Schlockow in his *Die Gesundheitspflege und medizinische Statistik beim preussischen Bergbau* (The Care of Health and the Medical Statistics in Prussian Mines) 1881, publishes statements made by Krieger that in the air of the mines there are many particles of dust with a diameter of less than one micron. Those in the sputum have a diameter of one to five microns. Discussing the sources of dust in coal mines, Schlockow differentiates between work in hard rock, in slate, and in the coal itself.

Hirt, in his work mentioned above, treats extensively of dust diseases, to which he devotes the entire first volume, *Die Staubinhalationskrankheiten und die von ihnen besonders heimgesuchten Gewerbe* (The Diseases Caused by Inhalation of Dust and the Especially Endangered Trades), Breslau, 1871. He gives much consideration to the forms of dust. He distinguishes according to the kind of dust and its form: (1) Metallic dust (a) with sharp and therefore injurious particles, as in iron and copper; (b) with blunt, roundish, noninjurious particles, as in lead and zinc. (2) Mineral dust (a) with sharp and therefore injurious particles, as in diamond, quartz, firestone, and marble; (b) with sharp and blunt particles, as in sandstone, clay, and limestone; (c) with blunt particles, as graphite. (3) Organic dust. He makes detailed proposals for the protection of workers, which we shall deal with later.

H. Eulenberg in his *Handbuch der Gewerbehygiene* (Handbook of Industrial Hygiene), Berlin, 1876, shows a much clearer insight into the connection between dust and injuries, pointing out that silicic acid, in whatever form it may appear, has its victims.

Even more important than these publications were those on animal experiments, among which we shall mention only the studies of von Ins and of Arnold. Von Ins made histological investigations, the results of which were later confirmed by other workers. He came to the conclusion that silicic acid remains in the lungs, but lime is dissolved and therefore is not damaging.[5] In the years 1878 to 1885 Arnold

---

[5] Adolf von Ins, "Experimentelle Studien über Kieselstaub Inhalation," *Archiv für experimentelle Pathologie*, V (1876), 169–94.

made experiments on various animals, some single experiments extending over a period of two and one-half years. He found that after fourteen days of persistent inhalation of quartz dust, there were to be found in the lungs nodules which later coalesced to larger indurations. These findings and others of interest were published in his book *Untersuchungen über Staubinhalation und Staubmetastase* (Research on the Inhalation of Dust and Dust Metastases), Leipzig, 1885.

A vast knowledge of dust on the lungs and silicosis had been accumulated by the year 1880. Then came Robert Koch's discovery of tubercule bacilli, to initiate the bacteriological era. At that time, particularly in Germany, the study of the effects of dust stopped. All cases were diagnosed as tuberculosis. The well-known Swiss hygienist, A. Vogt, mocked at all those "curiosities" of quartz lungs, coal lungs, and iron lungs, "all of which belong rather in a cabinet of curiosities than in industrial hygiene." [6]

Only a few authors defended the knowledge acquired regarding the effect of different kinds of dust on the lungs. Among these few were Füller and Schäfer, who, in the *Handbuch der Hygiene*, edited by Weyl in 1897, stressed the distinction between the condition of the lung caused by quartz and tuberculosis of the lung. Sommerfeld, in 1898, also held that the morphology of dust is important and that "it is sure that conditions which are still unknown or not yet sufficiently investigated play an important role, especially the chemical composition of the dust; for, if it were not so, why is sandstone more dangerous than granite and marble, although the latter two have sharper and more pointed particles?" [7] It is noteworthy that in the second

---

[6] A. Vogt, "Die allgemeine Sterblichkeit und die Sterblichkeit an Lungengeschwindsucht in den Berufsarten," *Zeitschrift für Schweizerische Statistik*, XXIII (1887), 279–80.

[7] Theodor Sommerfeld, *Handbuch der Gewerbekrankheiten* (Berlin, 1898), p. 19.

edition of Weyl's *Handbuch* (1913), the chapters on dust damages are far behind those of the first edition.

Very slowly the old knowledge as to the effect of dust on the lungs was revived—among others, in Germany by Koelsch in an article in the *Handwörterbuch der Sozialen Hygiene* (Dictionary of Public Health and Social Welfare), edited by Grotjahn and Kaup in 1912. Other authors tried to solve the question of the connection between dust and tuberculosis of the lungs by performing experiments on animals (Lubenau, 1907; Cesa Bianchi, 1913). In Great Britain, Arlidge in his book, *The Hygiene, Diseases and Mortality of Occupations* (London, 1892), treated extensively diseases caused by dust, distinguishing clearly between pneumoconiosis and tuberculosis. He also quotes Church, who analyzed the hardened lungs of potters and showed that in 100 parts of lung ash there were 47.78 parts of silica, 18.63 of alumina, and 5.5 of peroxide of iron.

The studies of conditions and mortality in mines and quarries continued and were often performed by departmental committees. The departmental committee appointed in 1892 to investigate metalliferous mines did not come to a clear decision as to the harmful effects of dust. Thomas M. Legge, however, writes in his annual report as Chief Inspector of Factories for 1900 that material obtained from two ganister miners by Dr. Robertshaw, Medical Officer of Health of the Stockbridge Union, and Dr. Maclaren, a certifying surgeon, and examined by the pathologist, F. W. Andrewes, proved beyond doubt the correctness of Dr. Greenhow's conclusions that inhalation of heavy angular particles of grit produces in the lungs iron gray nodules and the deposition of black mineral matter, and, further, that the development of tuberculous phthisis is intimately associated with—although independent of—the lesions produced by the dust.

We must also mention the investigation of slate mines

in Merionetshire (1892), and of Cornish mines (1902), as well as those of ganister mines made by Dr. Robertshaw and later (1901) by Legge.

The British Commonwealth, too, contributed to this research. In Western Australia, a royal commission investigated the mines (1904). Dr. Summons reported on miner's phthisis at Bendigo, Victoria (1907). Dr. Cumpston reported on pulmonary diseases among miners in Western Australia (1910), and a Royal Commission was appointed here the following year. Dr. Purdy reported on quartz mines at Waihi, New Zealand (1912).

But the revival and further development of the knowledge of dusted lungs, which was common before the discovery of the tuberculosis bacillus, stemmed from the British mortality statistics and from South Africa. It was to the credit of E. L. Collis that he pointed out, on the basis of governmental mortality statistics, that there are great differences between the age curve of "phthisis" mortality in the population as a whole and the curve of the "phthisis" mortality of men exposed to quartz dust. "Inhalation of mineral dusts which do not contain free silica tends to produce . . . respiratory diseases other than phthisis. Inhalation of mineral dusts which contain free silica is associated with an excess of phthisis, an excess which bears a direct relation to the amount of free silica present." Among the workers exposed to silica dust, he stated further, phthisis (1) occurs at a later age, (2) is associated with an increase of other respiratory diseases, and (3) shows a low degree of infectivity among female relatives.[8]

At about the same time, Haldane, a member of the Commission on Metalliferous Mines, summarized the results of his inquiries into miner's phthisis in Cornwall. He drew

---

[8] Edgar L. Collis, "The Effects of Dust in Producing Diseases of the Lungs" in *Seventeenth International Congress of Medicine* (London, 1913). Section XVIII, Hygiene and Preventive Medicine, Discussion No. I.

attention to the fact that all dust is not equally dangerous but that some—such as quartz dust from Transvaal gold mines, dust from Cornish mines, from ganister, and from some sandstone—is certainly injurious. "There is a great gap in our knowledge as regards these points," he stated. The commission tried to gather more material.

In South Africa, a report of the Government Mining Engineer (Transvaal) for the second half of the year 1901 stresses the high mortality among rock drillers. A Miner's Phthisis Commission was appointed in 1902, and its report was published in 1903. In 1907 a commission was named to inquire into the efficiency of existing regulations for mines and factories. In 1911 it was followed again by a Miner's Phthisis Commission, which published its report in 1912. In the same year, a new Miner's Phthisis Prevention Committee was called into existence. The general report of this committee, published in Pretoria in 1916, is the first complete modern discussion of the dust problem. It deals not only with mines and miners but with the problem as a whole and especially with silicosis, the clinical picture of which, by the use of X-rays, was first fully demonstrated by the men working in and for this committee. In this report the following parts are especially important: Appendix 6, "Silicosis ('Miner's Phthisis') on the Witwatersrand; Causation, Pathology, Symptoms and Radiographic Appearances" by A. H. Watt, L. G. Irvine, I. Pratt Johnson and W. Steuart; Appendix 7, "The Ashes of Silicotic Lungs" by J. McCrae; and Appendix 9, "Report on a Specimen of Dust from Silicotic Lungs" by J. Moir. Other appendixes discuss methods of sampling and counting dust and of dust prevention, to which we shall come back later. Other members of the committee who should be mentioned are W. Watkins-Pitchford and Robert Kotzé.

The report of the Committee stated that the essential factor in the development of chronic lung diseases among gold miners is the inhalation over long periods of the fine silicious dust generated by mining operations and suspended in the air. The majority of cases of early and intermediate silicosis are free from tubercular complications. Cases of silicosis may reach an advanced stage and may terminate in death without the intervention of tuberculosis. All mineral dusts are not equally dangerous; some do not appear to produce harmful results. The dusts which have been found to be dangerous are all characterized by the presence of uncombined crystalline silica. The silica should be in a state of minute subdivision in very fine, sharp-edged (and insoluble) particles. Seventy percent of the silica particles in the lungs are less than one micron in diameter; some of the others are as large as 8.5 microns. It was thus discovered that only quartz dust generates a specific disease, and that the smallest particles, about one micron in diameter, are the dangerous ones.

The research done by this Miner's Phthisis Prevention Committee, together with that of the previous committees, not only had an excellent effect on the miners' health in South Africa; it also helped to improve the dust trades of all civilized states. Later investigations failed to shake any of the committee's findings. Moreover, the flood of research set in motion by its reports has completed the clinical picture, and also our knowledge of the facts—important for industrial hygiene—concerning many dusty occupations. Before describing other studies, we wish to mention other South African investigations, especially those of A. Mavrogordato, "Studies in Experimental Silicosis and Other Pneumoconioses" (1922) and "Contributions to the Study of Miners' Phthisis" (1926); F. W. Simson and A. Sutherland-Strachan, "Silicosis and Tuberculosis" (1935). All of these appeared in the *Publications of the South African Institute for Medical Research,* Johannesburg. Further reports upon the work of the Miner's Phthisis

Medical Bureau brought out much interesting material. X-ray photographs now make possible a diagnosis on living persons which could not be made earlier with any certainty. The value of the X-ray picture has been established by numerous investigations. The first research work in Europe with the use of the X-ray, even before the South Africa research was known in Europe, was described by a Swiss author, Staub-Oetiker of Zürich, in 1916 in an article on pneumoconiosis of metal grinders.[9] Staub-Oetiker, together with Zangger, examined and took X-ray pictures of fifteen metal grinders. Later A. Böhme studied the pneumoconiosis of coal miners in the Ruhr district,[10] as did Patschkowski.[11] These two studies on dust lungs were the first in Germany in which X-ray pictures were used. The study of silicosis in other trades began in Germany much later with a study of metal grinders by Teleky, Lochtkemper, and Rosenthal-Deussen. Another study by the same authors followed (1932), and further articles on the subject, by Saupe, Koelsch, Kästle, Waetjen, Kalbfleisch, Gerstel, and others, were published in the volumes of the *Archiv für Gewerbepathologie.*

British contributions include government publications by E. L. Middleton and E. L. Macklin on grinders (1923), the first modern, complete research on this subject; by C. L. Sutherland and S. Bryson on the pottery industry (1926), sandstone workers (1929), and slate workers (1930); and by T. W. Wade on slate workers (1927). There were also the Milroy lectures by E. L. Middleton, published in 1936

---

[9] H. Staub-Oetiker, "Die Pneumokoniose der Metallschleifer," *Deutsches Archiv für klinische Medizin*, CXIX (1916), 469.

[10] A. Böhme, "Zur Kenntnis des Röntgenbildes der Lungen-Anthracosis," *Fortschritte auf dem Gebiet der Röntgenstrahlen*, XXIX (1922), 301; "Die Pneumokoniose der Bergarbeiter im Ruhrbezirk," *Fortschritte auf dem Gebiet der Röntgenstrahlen*, XXXIII (1925), 39.

[11] Patschkowski, "Ueber Pneumokoniosen bei den Bergarbeitern des rheinisch westfälischen Steinkohlenreviers," *Beitrage zur Klinik der Tuberkulose*, LVII (1923), 113.

under the title *Industrial Pulmonary Disease Due to the Inhalation of Dust*, and the studies in 1937 and 1938 by J. F. Bromley, A. E. Barclay, and collaborators, and by J. L. A. Grout. Last and most important were the Special Reports of the Medical Research Council, No. 243 (1942), No. 244 (1943), and No. 250 (1945), all carrying the title *Chronic Pulmonary Disease in South Wales Coal Miners.* The average annual incidence rate of pulmonary disease per 1,000 workers employed underground in the anthracite mines of South Wales was 5.23; in nonanthracite coal mines of South Wales, 0.99; and in coal mines in the remainder of Great Britain, 0.06. These very exact medical, environmental, and experimental studies try to explain these facts, which may be due to the higher dust concentration in anthracite mines or to differences of interaction of quartz with different types of coal, shales, and other strata.

In the Commonwealth, W. T. Nelson published his findings on miners in Western Australia in 1925 and 1926, and A. A. Ross and N. H. Shaw wrote about Australian foundries in 1943. Numerous studies by Charles Badham, Medical Officer of Industrial Hygiene, are published in the reports of the Director of Public Health of New South Wales.

The first investigations of health conditions in American mines were made by the United States Bureau of Mines together with the Public Health Service, and were conducted by A. J. Lanza and E. Higgins in the Joplin district, Missouri, in 1914 and 1915. Results were reported in the Bureau of Mines Technical Bulletin No. 105, 1915.[12] A later report by Lanza and S. B. Childs (Public Health Bulletin No. 85, 1917) concerned X-ray studies of silicosis in the United States. Investigations in lead mines followed

[12] A publication of a committee reporting on "life, working and health conditions" in the same district shows that in 1939 the same truly terrible circumstances prevailed.

in 1917 and in hard coal mines in 1935. R. R. Sayers, J. J. Bloomfield, J. M. Dallavalle, and others reported on the latter investigation in Public Health Service Bulletin No. 221. L. Greenburg and C. E. A. Winslow studied the dust hazard in grinding in 1920 (Public Health Report No. 35) and in sandblasting in 1931. Greenburg, with J. J. Bloomfield, reported on sandblasting and metallic abrasive blasting (1933). In addition to these studies, Public Health bulletins were issued concerning the health of workers in cement plants (No. 176, 1928), the granite industry (No. 187, 1929), and the mica and pegmatite industries (No. 250, 1940), as well as studies of other plants and mines.

In New York City a committee was organized to study silicosis among rock drillers, blasters, and excavators employed in the tunnels under the city. The results were published by Adelaide Ross Smith and by J. W. Fehnel (1929). The Department of Health of Pennsylvania studied the silica brick industry in 1939. From 1918 until his death in 1946, Leroy U. Gardner made very valuable experimental studies in the Saranac Laboratory on the effect of dust inhalation. Results were published in the *American Review of Tuberculosis* in 1920, 1922, 1923, and 1929 and in other periodicals.

Research on silicosis was furthered by the National Silicosis Conference of 1936 and by the establishment of the Air Hygiene Foundation.

The International Silicosis Conference in Johannesburg (1930) and Geneva (1938) gave a good survey of what had been achieved internationally in this field and pointed the way to further achievements.

The great volume of literature on silicosis published in those years is shown by *The Pneumoconiosis (Silicosis) Literature and Laws,* a bibliography compiled by G. G. Davis, E. M. Salmonsen, and J. L. Earlywine (Chicago,

1935–1937). In its three volumes are listed 2,005 titles. It is understandable that only the most important publications and a few others could be mentioned here.

In addition to quartz dust, other dusts affecting the lungs have been studied. Much discussed was the problem of whether or not silicates could cause damage similar to that of quartz. Some scientists held this to be true, but their arguments did not stand up under closer examination. There are only three kinds of silicates the dusts of which cause serious damage: asbestos, talc, and mica. Asbestos causes fatal lung changes which, however, show an entirely different picture from silicosis. The first case was reported by Montague Murray of the British Departmental Committee on Compensation for Industrial Diseases in 1906, the second by W. E. Cooke in 1924. The first more extensive publication on this subject is a report by E. R. A. Merewether (1930). In Germany, Fahr demonstrated to the Medical Society of Hamburg preparations and slides of an asbestos lung (1914). The next publications were those of W. Büttner-Wobst and Trillitzsch (1931) and of E. Krüger, Rostoski, and Saupe in the same year.

Talc, too, causes a fibrosis of the lungs with a roentgenographic appearance of granulations or nodulations which, to a certain degree, are similar to silicosis and have a disabling effect also. The first to describe such cases was W. C. Dreessen (1933). Then followed an article by Dreessen and Dallavalle (1935) and articles by F. W. Porro, J. R. Patton, and A. A. Hobbs (1942) and by W. Siegal, Adelaide Ross Smith, and L. Greenburg (1942).

Mica dust also produces lung changes. Fine, closely stippled markings appear in the X-ray picture, differing qualitatively from those of classical silicosis. This was first described by Dreessen, Dallavalle, and others (Public Health Bulletin No. 250, 1940). Finally we may point to the comprehensive work by Drinker and Hatch, *Industrial Dust* (New

York, 1936, 316 pages), which treats, primarily, the nonmedical aspects of the dust problem.

It is worth mentioning that a geologist, R. R. Jones, stated in an article published in the *Journal of Hygiene* (1933) that it is sericite, not silica, that causes the so-called silicosis. This article and others by the same author caused a sensation and for two years handicapped hygienic measures. But Jones' opinion could not be proved adequately and is now forgotten.

Based on these intensive studies, compensation for silicosis first was introduced in South Africa in 1912. Improved laws followed in 1914, 1916, and other years, up to the law of 1925. In Great Britain, silicosis was listed among the compensable diseases first for the refractories industries (1918, improved in 1925), then for the metal grinding industries (1927), for various other industries (1928), and for the sandstone industry (1929). In 1929 Germany introduced compensation for silicosis in sandstone work, grinding, china clay work, and mining, and extended it to all the trades in 1936. In the thirties, some states of the United States introduced such compensation—for example, New York in 1936. The peculiarities of this slowly developing chronic occupational disease necessitated special provisions in the compensation laws, some of which required periodical examinations of the workers.

CAISSON AND TUNNEL WORK

Diving without any protective device was performed in antiquity and is done even today by trained persons who dive for pearls. Aristotle, Arianus, and Pliny speak of vessels placed over the diver's head. Julius Caesar mentions a diving apparatus of leather tightened by resin. In 1511 Flavius Vegetius Renatus constructed a diving cloth that narrowed into a pipe, the opening of which was above the surface of the water.[13] We are told, but it has not been proved, that in Venice in the beginning of the 17th century a diving apparatus was used, into which air was blown by bellows. In 1691, Denis Papin suggested pressing air into the diving bell; Halley carried out the idea in 1716. This forcing of air either into the helmet of the individual diver or into the diving bell marked the beginning of a new period in diving. Such a diving apparatus seems to have been used by Becker in England in 1715. Before 1850 such devices were built by Siebe in England and by Heinke in the United States; they were called scaphander apparatus. Later, in 1865, a regulator for the compressed air was inserted between the pump and the lungs by Rouquayrol and Denayrouze.

Besides the diving apparatus into which air is blown, there are others which carry air with them. The oldest is that of J. A. Borellus (1682). In 1867 Sicard built one with a container filled with compressed air. Today some diving apparatuses contain compressed oxygen and are constructed in the same way as the self-containing breathing apparatus for rescue work.

The first diving bell is believed to have been constructed by Roger Bacon (1294). A Scotsman named George Sinclair (1665) and an American named William Philipps (1667) constructed diving bells for use in raising valuables such as jewelry and money from sunken ships. The Frenchman Coulomb and the Englishman Sweaton were the first to use a kind of diving bell for laying foundations (1778). The idea of an air lock for the transition from normal to increased air pressure was first proposed by Coulomb. In 1839 the French engineer Triger used a big iron tube filled with compressed air in building a shaft in coal mines.

[13] The following dates to 1900 are taken from the very extensive book: Heller, Mager, and Schrötter, *Luftdruckerkrankungen* (Vienna, 1900).

In 1850 the first caisson was constructed by Pfannmüller for bridge building in Mainz, Germany. Since that time the caisson has become the usual installation in bridge building. The Brooklyn Bridge and the Harlem Bridge in New York (1867) and the St. Louis Bridge across the Mississippi (1869) were the first bridges built in the United States in this way, according to Heller, Mager, and Schrötter.

Compressed air was used in the construction of a tunnel for the first time in Antwerp (1879) by L. Haskon; it was also used in building the first tunnel under the Hudson in New York (1879–1889), and from that time has been regularly employed in tunnel building.

Decompression Illness. Among the men diving without apparatus, bleeding from the nose and mouth was reported in the 16th century in divers in the Gulf of Mexico. Similar reports were made later by A. Herport (1666). Fatal cases also were described. Such bleeding is mentioned by authors up to the middle of the 19th century. But these symptoms have nothing to do with decompression; they may be caused by holding the breath. The first publication reporting an illness of divers who used apparatus with compressed air was written by Le Roy de Méricourt (1768) and was based on the observations of Denayrouze and Auble. This author described nervous disturbances, paraplegias, and deaths, and the publication attracted some attention. Other works on divers followed (Gal, 1872; Tetzis and Parissis, 1882). Catsaras (1890) wrote a comprehensive book with many observations on the illnesses of divers using various kinds of apparatus.

In occupations other than diving, Triger (1839) pointed to the dangers of working in compressed air. The first compressed air illness among miners was reported in 1845 in the mine at Douchy, which used Triger's method. B. Pol and T. J. Watelle in the *Annales d'hygiène publique* (1854) published facts on 63 such cases, most of them serious, two fatal. In bridge building there were similar reports from several parts of the world, and the frequency and seriousness of the illness increased as tunnels were constructed deeper below the surface of the water. Tunnel building, with its great number of workers, caused many cases of illness. Of the Hudson Tunnel, C. W. Moir reported in 1890: "When I came first to New York [in 1885] each month one man in 45–50 workers died. At the time when the work was done with a pressure of 35 pounds, the number of deaths rose to 7 men in 6 months." [14] Moir built a decompression chamber into which injured workers were brought. This improved conditions, and the number of fatal and serious cases diminished. But in all construction of bridges and tunnels the number of injured workers remained very high.

In 1854 Pol and Watelle already had pointed to the fact that it was not the increased pressure, not the stay in high pressure that caused the damage, but the transition from high to normal air pressure. They thought that the illness might be caused by the congestion of the blood in the internal organs as a consequence of the increased pressure on the surface of the body, a visceral congestion which develops serious symptoms after decompression. Many other theories followed.

Boyle as early as 1672 and later van Musschenbroeck in 1755 had noted the liberation of gases from the blood into the blood vessels of animals which were brought into air having low barometric pressure. Hoppe-Seyler made the same observation in 1857 and used it as an explanation of the sickness reported by Watelle. But the correct explanation was given by the Frenchman Paul Bert in 1872 after many experiments and investigations. He came to the conclusion that all these accidents in decompression were "the consequence of the liberation of bubbles of nitrogen in the

[14] Heller, Mager, and Schrötter, *op. cit.*, p. 364.

blood and even in the tissues when the compression had continued for a long time." [15] In 1878 Heiberg demonstrated by autopsies the presence of gas bubbles in the blood vessels of workers in compressed air. In a higher atmospheric pressure, the blood and the tissues take up more air. The oxygen is chemically bound, but the nitrogen is only absorbed and becomes free when the pressure decreases. This process develops more intensively, the greater the amount of gas absorbed and the quicker the compression decreases. The amount absorbed depends on the height of pressure as well as on the length of time during which the worker is subjected to the high pressure. The danger spot is decompression, which, unless performed very slowly, causes nitrogen bubbles to be liberated in the tissues, producing various symptoms.

It should be mentioned that the clinical picture has—in contrast to all the manifold acute forms—a chronic form also, which produces changes in the skeleton, especially in the hips. This form was first described by Plate (Germany) and Bassoe (United States) in 1911 but then was nearly forgotten until it was again observed by the Germans Christ (1934), Frank (1935), and F. Jaeger (1937), by the Americans S. C. Kalstrom, C. C. Burton, and D. B. Phemister (1939), and then by others.

Surveying the numerous observations and research studies made by Snell (1896), Heller, Mager, and Schrötter (1900), Boycott, Damant, and Haldane (1908), Japp (1909), Keays (1909), Waller (1910), Bornstein (1910 and 1912), Koelsch (1915), E. Levy (1922), and Singstad (1936 and 1939), we can say that there is now a clear picture of all the conditions and their importance in causing caisson sickness and also of the methods of avoiding it. The best working method—the right proportion between pressure, working time, and

[15] *Ibid.*, p. 394.

decompression time—seems to be that given in a table issued by the British Institution of Civil Engineers' Committee on Regulations for the Guidance of Engineers and Contractors for Work Carried Out in Compressed Air (1936). It recommends decompression by stages (proposed by Haldane) and the use of certain decompression periods graduated in accordance with the extent of the compression and the time spent working in it.

Besides this, good ventilation of the workroom is very important (Lipkowic, 1937), and equally necessary is thorough medical supervision of the workers, as well as the treatment of sick persons in the recompression chamber. The use of oxygen during the decompression has been recommended (Jones, Crosson, and others, 1940), but it has not found further approval at present. Helium has proved very useful in deep-sea diving, where conditions differ in certain respects (Behnke, 1942).

Legislation for the purpose of protecting workers in compressed air began relatively late. First the Netherlands published such a law on May 22, 1905, and regulations on January 26, 1907. France followed on December 15, 1908, New York in 1909, New Jersey and Pennsylvania in 1912. Germany published its regulations in June, 1920, which were changed in May, 1929, and in 1935. Great Britain has at present no governmental regulation of caisson work, seemingly because it cannot be included in the Factory Acts. The report of the Institution of Civil Engineers is mentioned above. It should be added that laws concerning sponge divers were passed by Greece in 1910 and by Italy in 1913. The British Admiralty issued instructions for divers in 1907, the German Admiralty in 1910, both of which have been improved several times.

In most countries, regulations for work in compressed air contain the above-mentioned precautions, especially where the pressure surpasses a certain height (1.3 kg. per square cm.). They also require slow decompression, short working time (especially in higher pressure), presence of a recompression lock, provisions for first aid, medical pre-examination, and periodical examination of the workers. There are, however, differences between the countries. In Europe, longer working hours and consequently longer decompression periods are preferred. In the United States, working hours and decompression periods are generally shorter. In some American states the decompression time is definitely too short.

Among the many publications concerning research and careful practical studies performed during the last decades, the following deserve mention. Books describing the clinical course, and control of caisson sickness published at the beginning of this period include E. H. Snell, *Compressed Air Illness* (London, 1896), the standard work, R. Heller, W. Mager, and H. von Schrötter, *Luftdruckerkrankungen* (2 vols., Vienna); and L. Hill, *Caisson Sickness* (London, 1912, 255 pages). In the later decades no book treating the whole question has been published, but there have been some very valuable booklets, notably that of the British Institution of Civil Engineers, *Report of the Committee on Regulations* (London, 1936). The great tunnel works in the United States, especially in New York, afforded opportunity for valuable studies: F. L. Keays, *Compressed Air Illness, a Report of 3,692 Cases* (New York, 1909); E. Levy, *Compressed Air Illness and Its Engineering Importance* (U.S. Bureau of Mines Technical Paper No. 285, 1922). In recent years the United States Navy has made valuable experiments in diving, reported in articles by A. R. Behnke and his co-workers in the *American Journal of Physiology*, CXIV (1935–1936), and the United States Naval Medical Bulletin No. 40, 1942.

We do not wish to deal here with the special problems

SPECIAL PROBLEMS 215

either of diving or of aeronautic service, although the latter may become one of the most interesting problems of occupational hygiene in the future.

SKIN DISEASES

All the authors who deal with occupational diseases mention skin diseases.[16] Paracelsus and Agricola discuss them briefly; Ramazzini (1700) mentions skin diseases of lime workers, fullers, cleaning women, salt workers, and smutters (the latter caused by little worms), and also refers to ulcers of the legs in people whose work requires them to stand. He states that bakers and millers usually have lice. Ackermann confirmed this in 1780; in writing of the "illnesses of dusty trades" he cites skin diseases caused by dust.

When at the end of the 18th century the fundamental basis of a scientific dermatology was created, increasing attention was given to occupational skin diseases, especially by English and French authors. Robert Willan gave an extensive description of baker's itch and of the skin diseases of cleaning women. He and his pupil Thomas Bateman (1815) described skin diseases in many other trades, among them mason's itch. Many such dermatoses also were described by J. J. Alibert between 1806 and 1827 and by the German, Johann Frank (1815). Halfort describes (1845) the skin changes in arsenic workers due to Schweinfurt green. This is also treated by later authors: Blandet (1845), Chevallier (1847 to 1859), Imbert-Gourbeyre (1857), and others. Potton (1851–1852) was the first to describe the "mal de bassine," the dermatitis of workers handling cocoons of silkworms. Armieux (1853) and Lunel (1859) were the first to describe tanners' "rosignol" or "pigeon's eye." Reynal

---

[16] Data before the year 1870 is drawn from the chapter by I. Fischer, "Geschichte der Gewerbedermatosen" (History of Occupational Dermatoses) in *Die Schädigungen der Haut durch Beruf und gewerbliche Arbeit* (Injuries to the Skin by Trade and Occupation), Leipzig, 1922, Vol. I.

(1857–1858) studied herpes tonsurans, transferred from animals to men. M. Vernois' comprehensive study of the hands of workers was published in the *Annales d'hygiène* (1862). In the same journal, Becourt and Chevallier discussed bichromate ulcerations (1863). Many other publications followed but, as we recognize from those just mentioned, French authors were the first to study this field.

Statistics of the German, Cless (1842), and the Dane, Hannover (1861), showed that "no trade has so many and such obstinate skin diseases as the bakers." A combination of factors resulted in this frequency: the humidity, the flour, and especially the poor hygienic conditions of workshops and dwellings. As late as 1900 to 1904 Sternberg, chief physician of the Viennese Sickness Funds, reported that 5.5 percent of the baker journeymen were treated for scabies yearly.

In the middle of the 19th century, Hebra and Kaposi conducted certain etiologic studies of dermatoses that made it possible to differentiate between skin diseases caused by external influences and those caused by inherent peculiarities (for example, psoriasis) or by systemic illness. The books dealing with occupational diseases and industrial hygiene of this period also discuss skin diseases—A. Layet (1875), Hermann Eulenberg (1876), and Ludwig Hirt in a special chapter in the last volume of his four-volume work. Arlidge (1892) treats of skin diseases in a very short chapter but mentions some that are peculiar to individual trades. In *Dangerous Trades*, edited by Thomas Oliver (1902), skin diseases are more extensively treated. In the chapter "Dust as a Cause of Occupational Diseases" there are sections on skin diseases of flaxworkers and on diseases of nails in furriers. In the *Handbuch der Arbeiterkrankheiten* (Handbook of Workers' Diseases), edited by Theodor Weyl (1908), there is a chapter on occupational skin diseases written by A. Blaschko. During the last decades not only many articles but voluminous books on occupational skin diseases have been published.

A great advance was made in our knowledge of occupational skin diseases, especially occupational eczemas, when Pirquet's discovery of "allergy"—that is, "acquired specific alteration in the capacity to react which occurs in living organisms upon exposure to certain substances"—was applied to occupational eczema. The skin acquires a hypersensitivity to a certain substance after repeated or prolonged contact with it, and the eczema develops. That explains why occupational eczema does not develop immediately after a short contact, but after varying periods of time, sometimes after years of contact, and why it recurs after a fresh contact with this substance. Jadassohn and Bloch (1923–1925) worked out methods of testing the skin for its sensitivity to various substances. These methods have been developed further by many authors (R. L. Mayer, H. Jaeger, Hans Stauffer, Wilhelm Berger, and Sulzberger). They give us not only a deeper theoretical insight into the nature of most of the occupational eczemas but also permit us to determine exactly what causes eczema in a person: whether it is a specific substance, its admixtures, or impurities, present within it. This determination is very important in avoiding further relapses and in the question of compensation. In some countries all occupational skin diseases whose origin by occupation has been established must be compensated, in other countries only those caused by certain substances. Investigations during the last few years have attempted to distinguish the occupational eczemas of allergic nature from those having other origins, such as wear and tear (Bering). Of course we do not have at all in mind the acute damages to the skin caused by corroding substances, burns, and so forth.

As to the proper hygiene and the methods of avoiding skin diseases when contact with an injurious substance is involved, recommendations for cleanliness, protective ointments, and rubber gloves have always been made. Except for the fabrication of thinner, more resistant, and cheaper rubber gloves and better ointments, no true progress is being made at present. However, skin tests that make it possible to determine exactly which are the harmful substances show how best to avoid such danger.

As an illustration we cite the historical changes in the skin diseases of one occupation. I mentioned above the skin diseases of bakers, among which pediculosis and scabies played an important role. With the improvement of the especially bad housing and working conditions of bakers, these diseases have disappeared during the last decades, while the incidence of other skin diseases frequent among bakers has declined. R. Prosser White wrote in 1928 that eczema occurred no more frequently among bakers than among other people. Later he wrote "Eruptions certainly and solely caused by the materials bakers handle in their work are probably exceptional." And in a footnote: "No positive experimental evidence has so far been forthcoming. . . . No specific descriptions or analysis of distinguishing features are available. In former days it may have been originated from the presence of the *Acarus farinae*." [17] Later research was done by A. C. Parson under the sponsorship of the British Bakers' Union; he found an increased frequency of eczema in bakers of certain districts only. The same was true of certain districts in Germany. The number of bakers suffering from eczema in Germany as a whole rose from 0.27 percent in 1923 to 1.31 percent in 1930. These figures varied in different districts. At this time Josef Kulman worked out methods for easy detection of certain chemicals which have been used in recent years for "improving" and "bleaching" flour, processes used in

[17] R. Prosser White, *The Dermatergoses or Occupational Affections of the Skin* (3rd ed.; London, 1928), p. 314.

large mills in several European countries and in the United States. Using these tests, Zitzke and Teleky (1932) proved that in the Rhineland the majority of cases of eczema in bakers were caused by ammonium persulfate, used as a flour improver. That was confirmed later by other authors. The use of persulfates in smaller amounts, or their replacement by other chemicals, was therefore the best method to prevent the majority of cases of baker's eczema.

Attempts to select for certain trades persons who are not apt to become sensitive to the eczematogenous substances used has not succeeded as yet. Nor has success followed in most cases the attempt to desensitize persons who have become hypersensitive. But it is to be hoped that in the future these efforts may succeed and that the greatest number of cases of occupational eczema may thus be prevented or cured.

The following important books concerning occupational skin diseases have been published in the last decades: R. Prosser White, *Occupational Affections of the Skin, Their Prevention and Treatment,* 1st ed., London, 1915; 4th ed. under the title *The Dermatergoses,* London, 1934, 716 pages; K. Ullmann, M. Oppenheim, and J. H. Rille, *Die Schädigungen der Haut durch Beruf und gewerbliche Arbeit* (The Injuries to the Skin by Profession and Occupation), Leipzig, 1922–1926, 3 vols.; R. L. Mayer, *Das Gewerbeeczem* (Occupational Eczema), Berlin, 1930, 98 pages; Otto Sachs, "Gewerbekrankheiten der Haut" (Occupational Diseases of the Skin), in the *Handbuch der Haut und Geschlechtskrankheiten* (Hand-book of the Diseases of the Skin and Venereal Diseases), ed. by J. Jadassohn, Berlin, 1930, Vol. I, 200 pages; E. Bering and E. Zitzke, *Berufliche Hautkrankheiten* (Occupational Diseases of the Skin), Leipzig, 1935, 264 pages; Louis Schwartz and L. Tulipan, *Occupational Diseases of the Skin,* Philadelphia, 1939, 799 pages; 2d. ed., with S. M. Peck, 1947, 964 pages.

11

Mines and Miners

THE HYGIENE of mines and miners and the position of the mining industry in economy and legislation is entirely different from that of other industries. The miners are—at least on the Continent and apparently in the United States also—such a specialized group of working people and the dangers and injuries to which they are exposed are so common to nearly all of them that a special treatment of the history of the hygiene of mining is necessary, the more so since the problems differ very much from those of factory hygiene, and consequently mining legislation differs from factory legislation.

LEGISLATION

The position of miners seems to have varied very much in different countries. In the early Middle Ages in all European countries, it was the state, not the landowner, that owned the minerals below the surface of the earth. According to Arndt, this system—which goes back to the time of the Phoenicians and Romans—is still prevalent on the Continent. The sovereign grants a person or a society the right to raise ore, coal, and other minerals with certain reservations, among which, as a rule, is the obligation to exploit the mines according to his regulations and under his supervision. The mineowner is independent of the landowner to whom the surface over the mine belongs.

In Germany, Emperor Karl IV ceded to the rulers of the individual countries his right to the mines in the "Goldenen Bulle" (1356). That was not really a change from the old system. The principalities issued *Bergordnungen* (rules for mines), some very early, most of them in the 16th century or later. At first these sovereigns gave the right to mine to the *Gewerken*, a small number of men working as miners themselves. Of course, they had to pay certain imposts and fulfill certain obligations. Gradually the mines became deeper, and more capital was needed for their operation. So these *Gewerken* admitted persons who were not miners but who possessed the necessary money. Slowly societies replaced the former *Gewerken*, and the position of the miners deteriorated. On the other hand, the miners were skilled workers and the mineowners needed them. So their position was better than that of many other workers. They were the first group for which a kind of accident compensation was created (Joachimsthal, now in Czechoslovakia, 1541; Cologne, 1669). Skilled miners were wanted by the sovereigns of neighboring countries, and thus German miners appeared in the mines of Bohemia, Poland, and Hungary, enjoying certain privileges (freedom from taxes and from military service). Other German skilled workmen also were called into these countries to increase the industrial activity and the trades. The descendants of these miners, artisans, and craftsmen continued to live here, most of them working in the same trades till 1946, when they were driven out in accordance with the agreement of the Allies at Potsdam.

The societies described above often developed into stock companies; and as their power increased, the position of the miners deteriorated even more. However, the mines had to be run according to the regulations of the sovereign and his officials ("Directionsprincip"). The mine officials were often appointed by the government, and many details of management could be performed only in agreement with the governmental officials. For instance, in some districts dismissal of workers or their transfer to other mines was under governmental control. But that does not mean at all that hygienic installations and accident prevention were always good or even sufficient.

The Saxon mine law of 1851 and the Prussian mine laws of 1851 and 1865 removed all this "tutelage." The mine authorities from then on had to supervise only the protection of the mines, of the surface, and of the lives and health of the workers. The relation of the miners to the mine and its proprietor then became the same as that of industrial workers to their employers. Many strikes and struggles over the *Knappschaftskassen* (the funds for accidents, illness, and death benefits) broke out, and of course also over wages.

After the abolition of the "Directionsprincip," the number of accidents rose. In the period from 1841 to 1850 there were 1,680 accidents per 1,000 miners a year in the mines of Prussia, and in the period from 1866 to 1875 this rate had risen to 2,471. The powers of the mine authorities which had been granted them for centuries now were restricted to certain tasks, and the officials had to adapt themselves to the new circumstances and find the way for further protection of mines and miners.

The Minister and the "Oberbergämter" (superior mine authorities for larger districts) have the right to make certain rules. A mine is allowed to be run only in accordance with a plan approved by the authority. The manager must be authorized by the government. A worker may work as a hewer only after three years of mine experience and after having passed an examination. There are many technical regulations concerning safety, ventilation, wetting and dusting, safety lamps, blasting explosives, rescue apparatus (one for every 200 workers), and shower baths. As to the latter, it may be mentioned that already in 1893, 95.7 percent of the 138,440 coal miners of the district of Dortmund (Ruhr District) had bathing accommodations at the exit of the mines.

The regulations are very extensive and cover every detail. For example, the regulations of the year 1911 for coal mines in the Ruhr District, a district with many dangerous mines, comprise 374 paragraphs (100 pages). Other regulations concern the use of the cage (1927), the work of hewers (1925), and the use of stone dust against explosions (only approved dust has been allowed since 1925). A collection of the most important mine regulations for this district (those for some quarries included) was published by Schlattmann, a highly qualified governmental mine official, in a book of 332 pages.

Most of the coal mines—those which produce 96 percent of Germany's coal—and the greatest part of the other mines are located in Prussia. They are supervised by the Department of Mines in the Prussian Ministry of Commerce. There are five "Oberbergamtsbezirke" (Superior Mine Authorities) together with 546 employees (clerks not included), all of whom are well-trained mining engineers. Since 1921 there has been a special "Grubensicherheitsamt" (Mine Safety Office), which published yearly reports.

In the England of the early Middle Ages we find the same ideas that prevailed on the Continent regarding the possession of mines and resources below the surface of the earth. But King John was—some authors say—compelled by the Great Charter (1215) and the Forest Charter to resign his rights to all the mines. This is not clearly expressed in these two documents, but is to be found in the Act of Henry IV (1404) and William and Mary (1688). Since that time the minerals below the surface have belonged to the landowner. That brought the miners who did not work on their own ground into dependence on the landowners. In many districts, miners lived on a very low level, especially in northern England. R. N. Boyd quotes an author of the 17th century as saying that the coal miners in these districts are "regarded by their neighbors as outcasts and almost at war with Society." J. U. Nef points out that the Scotch miners were, together with their wives and children, the slaves of their employers. In 1662, two thousand coal miners presented a petition to the King in which they complained of the wrongs done them by the mine owners and overmen and asked for more ventilation shafts. About 1690 the coal miners began to set up funds for illness and unemployment, but the employers tried to get these funds under their own control. In the Scotch Habeas Corpus Act of 1701 it was provided that the "Act is in no way to be extended to colliers or salters." It was not before 1775 that these workers obtained equal rights.

The first legislation on mines, the Malicious Injuries Act (1769) and the Act for the Security of Collieries and Mines and for the Better Regulation of Colliers and Miners (1800), provided for the punishment of miners who damaged machines for draining collieries and mines, or who did not observe the contract, or who stole coal.[1]

The first explosions were reported in 1675 and 1677, then in 1705 and later. In the year 1812 an explosion in a coal mine killed 92 miners. After that disaster the Sunderland Society was founded in 1813. This group promoted the hygiene of mine workers, with the help of John Buddle, a great mineowner, and Sir Humphry Davy, inventor of the Davy safety lamp. Buddle described, recommended, and built ventilation equipment.

The Act of August 10, 1842, initiated by Lord Ashley, prohibited the underground work of women and also of children below the age of ten years and gave the Secretary of State the right to appoint inspectors who should visit and inspect the mines.

The first inspector was Tremenheere. He did much, but "there is no evidence that he ever made any routine under-

---

[1] Sources used here include: R. N. Boyd, *Coal Mines Inspection; Its History and Results* (London, 1879); J. U. Nef, *The Rise of the British Coal Industry* (London, 1931); Mining Association of Great Britain, *Historical Review of Coal Mining* (London, 1924); E. L. Collis and Major Greenwood, *The Health of the Industrial Worker* (London, 1921); and G. Rosen, *The History of Miners' Diseases* (New York, 1943).

ground inspection, and apparently it was not expected of him" (Rosen). The miners complained that the law was frequently infringed upon. The government charged Professor Lyell and Michael Faraday and later two others with investigating the causes of mine disasters and explosions. Lyell and Faraday recommended ventilation by means of pipes and tubes.

The miners, organized in a National Union (1841–1847), finally brought pressure upon the government. Indeed, the Act of 1850 for the inspection of coal mines called upon owners to give notice of fatal accidents and to keep accurate plans of their mines, and also settled the principle of underground inspection. But the government made the inspectors "really little more than scientific assistants to the coroner" (Boyd). However, the act had at least the effect of bringing about a thorough investigation of accidents. Six inspectors were appointed, but mine accidents did not decrease; in the period between November 1, 1850, and December 1, 1852, 2,040 miners died in colliery accidents.

When the Act of 1850 was replaced by that of 1855, adequate ventilation by mechanical means was required, sufficient to render noxious gases harmless. The act laid down a standard of safety by ordering the observance of seven "general rules" which were to be applied to all mines and by requiring each individual colliery to enforce its own code of special rules with the approval of the Home Office. This latter regulation was as ineffective as a similar one covering factories. The Act of 1862 asked for two shafts in every mine. In 1863 a mine disaster in which 204 miners suffocated caused strong agitation among the miners, and the Miners' National Association was founded with the purpose of furthering the interests of the miners in legislation, inspection, and accident prevention (Boyd).

In 1872 a Mines (Coal) Regulation Act and a Metalliferous Mines Act reinforced the provisions dealing with ventilation and safety. The second asked for accommodations for drying the miners' clothes. It required, further, that every mine have a manager holding a certificate of competency obtained in a state examination.

But again and again disasters and explosions occurred. A Royal Commission appointed in 1879 issued a report in 1886 pointing to the danger of coal dust and recommending the use of high powered explosives, the testing of safety lamps, and the installation of rescue stations. An Act of 1887 reinforced some of the existent regulations but did not introduce any of the measures recommended by the commission. Slowly the duty of the government to enforce safe management of the mines was recognized. D. Morrah, from whose chapter in the *Historical Review* some of these facts are taken, thinks that this recognition came just after 1872, but it seems to me that it was later—after 1900.

An Act of 1910 authorized the Secretary of State to require and make arrangements for rescue work and to train men for it.

By the Act of 1911, certificates were required also for assistant managers, firemen, and other mine employees. The act also gave directions on drilling by mechanical power, which should not be done in highly silicious rock unless a water jet, spray, or other means equally effective in preventing the escape of dust into the air were employed. Facilities for bathing and for drying clothes were required to be installed where the majority of miners wished it. The Secretary of State obtained the right to amend or to add regulations to this code of safety rules. In the following years, he issued regulations concerning rescue work, washing and drying facilities (1913); work on ganister (1920), and explosives (1913, 1915, 1918, and 1919). Orders concerning safety lamps were issued in great numbers between 1913 and 1919.

An Act of 1920 created a special Mines Department with

a Secretary of Mines in the Board of Trade and with an advisory committee of employers and employees. A fund, the Miners' Welfare Fund, was created to be used for purposes connected with the social well-being, the recreation, and the living conditions of mine workers. It was financed by a tax of one penny per ton of the output. After 1934 this was reduced to a halfpenny per ton, but was raised again to a penny in 1939. The fund was administered by a commission of employers, employees, and governmental officials.

Morrah wrote in 1924: "We have reached a time when official control extends to every detail of mining routine. Is the condition of our mines really better because it is more highly regulated? We need not be apprehensive for the answer. The mining industry once so hazardous and speculative, has become the mainstay of our economy." [2] Since that time, legislation and the formulation of rules have progressed in the same direction.

The law on the welfare fund was extended in 1931 and 1934. The administrative commission of the fund publishes annual reports. It promoted the installation of baths at mines. In 1939 there were 313 bathing installations for approximately 400,000 men. Canteens and first aid rooms have been installed and other welfare work has been initiated: convalescent homes, playing grounds, and so forth.

The regulations concerning the work of women and the working time of young men were made stricter in 1935, 1936, and 1938. In 1938, far-reaching rights were given to the government regarding the consolidation of coal mines, owing to the fact that many small mines were in financial difficulties. Extensive "Coal Mines (Ventilation) General Regulations" were published on May 17, 1947.

[2] Dermat Morrah, "A Historical Outline of Coal Mining Regulations" in Mining Association of Great Britain, *Historical Review of Coal Mining* (London, 1924), p. 319.

The nationalization of coal mines has now been carried out.

The governmental office in Great Britain that deals with mines is the Mines Department. After public hearings it issues rules and regulations, which must be submitted to Parliament. For investigation there is the Safety in Mines Research Board. The Chief Inspector of Mines has two deputy chief inspectors, one medical inspector (Dr. Fisher), two electrical inspectors, and one for special duties. The Kingdom is divided into eight districts, each headed by an inspector in charge, one or two senior, and several junior and subinspectors.

The Mines Department works not only with its own staff but has the cooperation of the organizations of miners and mineowners as well. In addition, a Royal Commission on Safety in Coal Mines and the Safety in Mines Research Board (previously mentioned) were created. The Board was appointed in 1921. Among its members are J. B. S. Haldane, Director of the Mining Research Laboratory of Birmingham University, J. W. Thorpe, R. V. Wheeler, and S. M. Dixon. Several laboratories and research stations are associated with the Board. On some problems it cooperates with the Medical Research Council, which handles most of the questions of public health. Between 1921 and 1938, the Board published 102 reports and booklets, one "information paper," and seven leaflets. (Some of the institutions named here will probably be changed in consequence of the nationalization of mines.)

In the United States the supervision of mines is the task of the individual states. In 1944, six states had an output of more than $100,000,000 worth of coal and eleven states produced between $50,000,000 and $99,000,000 worth. The states having the largest output in 1942 were Pennsylvania (141,600,000 net tons), West Virginia (159,600,000 net tons), then—far behind—Illinois (72,400,000), Kentucky (61,900,000) and, still further behind, Ohio (32,000,-000), Indiana (25,000,000), and others.

The total number of mine inspectors in all the states was 279 in 1942; 56 were in Pennyslvania and 50 in West Virginia. States with a large coal output should be expected to do their best to maintain the safety of mines. But E. A. Wieck points out in his *Preventing Fatal Explosions in Coal Mines* (1942) that a continuous battle is being waged in the state legislatures between leaders of the miners' union, who ask for better legislation, and the operators, who oppose it. Generally the mine operators have strong influence, so the state mining codes are a mixture of good and bad, instead of being scientifically drawn up. The mine inspectors must pass an examination in every state, but almost all of them are appointed for political reasons and change with the political administration. In five states—among them Michigan, Oklahoma, and Idaho—the mine inspectors are elected by the people for two to four years, according to Andrews (1938). Therefore many good inspectors, wishing to do their best, are hampered by politics.

An excellent federal institution is the United States Bureau of Mines—as a research institution. It was founded in 1910 as a section of the Department of the Interior, was transferred to the Department of Commerce in 1925, but returned to the Department of the Interior in 1934. The purpose of this Bureau is to make "inquiries and scientific and technologic investigations" (1913). Later the study of occupational diseases was added to its tasks. But the law declared expressly that the Bureau had no right or authority to inspect or supervise mines in any state.

The first mining experiments were performed and mine safety stations provided in 1915. From that time on, the Bureau developed an extensive activity and performed a tremendous amount of scientific work. It publishes major investigations and extensive studies in the form of bulletins.

Its "Technical Papers" offer reports on the progress of investigations. It issues "Economic Papers," "Minerals Yearbooks," handbooks for use in safety measures and mine rescue, schedules discussing standards for permissibility of apparatus and materials, "Miners' Circulars" giving miners and mine officials advice in safety, and finally annual reports in which the Director reviews the activities of the Bureau. There is a list of publications in which every publication issued by the Bureau between July, 1937, and December, 1942, is treated in three to seven lines. To this 98-page booklet there is a yearly supplement of about 30 pages. (See also Chapter 8, p. 163.)

The Bureau did much practical work in this field by examining methods and devices and stating the requirements which had to be fulfilled to make them effective and safe. When certain conditions were met, it declared methods "permissible" and gave apparatus the "approval label." It was the first governmental office in the world which tested and approved (or did not approve) industrial masks, respiratory devices, and rescue apparatus. It approved safety lamps which met the requirements of the Bureau, tested explosives and blasting units, declared some "permissible," and gave advice as to their use. Investigating rock-dusting and wetting, the Bureau came to the conclusion that they supplement each other in the prevention of coal dust explosions, and inaugurated a definite trend toward the combined use of water and rock dust. Removal of coal dust (cleaning) cannot entirely prevent disasters; therefore the Bureau recommended spraying water during machine cutting up to a distance of forty feet, but generalized rock-dusting in all coal mines except anthracite mines, supplementing it by rock dust barriers.

This great Bureau has a large and expert staff. (To mention only a few, among its most competent men are Director R. R. Sayers, H. H. Schrenk, D. Harrington, and G.

W. Grove.) However, *it does not have the authority to issue any orders or rules;* it may only give advice. Until comparatively recently it did not even have the right to enter any mine without the consent of the owner or operator. Finally Public Law 48 (April 26, 1937) created a commission for bituminous coal mines, which was charged with investigating—among other matters—the safe operation of mines in order to reduce working hazards to a minimum. These investigations were to be made in collaboration with the United States Bureau of Mines. It was Public Law 49 (May 7, 1941) which authorized the Secretary of the Interior, in cooperation with the Bureau of Mines or other safety agencies, to make annual or necessary inspections and investigations in coal mines whose products regularly enter or substantially affect interstate commerce, in order to obtain information on health, safety, causes of accidents, and occupational diseases therein. Inspections may also be made after accidents. The law also provides for the appointment of coal mine inspectors, who must have had five years of practical experience, the practical training and experience of a coal mine engineer, safety inspector, or mine foreman. Up to June, 1944, 2,400 mines have been inspected once, and 1,400 two or three times. But J. J. Forbes, a representative of the Bureau of Mines, told a Senate committee that prior to the safety code (July, 1946) only about a fourth of the recommendations of the Federal mine inspectors were carried out, most of them by the more progressive companies.

In the chapter about Workers Unions, especially the United Mine Workers of America, we described how the Federal Mine Safety Code originated from an agreement between the Union and the Federal Coal Mine Administrator and how after a disaster the rights of the Federal mine inspectors were augmented. The Bureau of Mines and its inspectors have no right based on law to issue rules or orders or to arrange anything in the mines. That is within the rights of the individual states, activity of which is very different and sometimes doubtful.

We pointed in the above-mentioned chapter to the weakness of a guarantee of safety by agreement. J. J. Forbes said before the committee: "Since the coal mines have been in government possession I know that enforcement of compliance with a safety code is the only means by which the coal mines of the Nation can be made safe. . . . I believe that the Bureau of Mines should be given authority to enforce the provisions of a reasonable code."

I myself am of the opinion that this is the only efficient way to make American mines as safe as those of Europe. May we hope that not too many disasters are necessary before the impediment to such an authorization has been overcome.

THE DANGERS AND THEIR CONTROL

We have seen that in the trades the change from little workshops to larger, from handwork to machine work, and the development of capitalism made special protection of the worker necessary. What were the circumstances which necessitated a thorough protection of miners? We are told that even in antiquity there were large and deep mines. The silver mines in Laurion (Greece) were more than 400 feet deep. In Spain there were large mines in Roman times. But all that was changed with the decay of antiquity. In the Middle Ages and at the beginning of modern times, mines were, as a rule, only 80 feet deep (Agricola, 1556). Hue (*Die Bergarbeiter*, 1910–1913) thinks that a century later the deepest shaft went down as far as 800 feet below the surface of the earth; the reports of mines in Kitzbühel (Tirol) and Kuttenberg (Bohemia) which were said to be 3,000 feet in depth, he considers unreliable. At the beginning of the 18th century the average pit depth in Cumberland, where the collieries were deeper than elsewhere in England, was about 190 feet. At the end of the century a depth of 900 feet was attained; about 1830 one of 1,500 feet was reached. Today the average depth of English mines is 900 feet, the deepest being about 3,300 feet. The greater depths necessitate complicated installations, strengthened and improved, as in the case of pumps to keep water out of the mines. We shall discuss here only certain installations having a special effect on health and safety.

The depth also makes mechanical installations necessary to bring miners to their working places. As late as 1871 Barham was still able to say that the Cornish miners had "to ascend a thousand feet or more by ladders, the universal mode thirty years ago, but now superseded by machinery in about a dozen of the most important mines." [3] Rosen states, "The introduction of the metal cage (*Förderkorb*) towards the middle of the century and its increasing adoption . . . led to the disappearance of the more primitive methods." [4] These primitive methods consisted of—in addition to climbing ladders—the use of a basket, or a rope with a primitive mechanism, called a "man-engine" (*Fahrkunst*). According to Meissner, in the years 1883 to 1892 in Prussia, the following number of fatal accidents occurred per thousand miners: [5]

Those using the metal cage (*Förderkorb*)	0.043
Those using the ladders (*Fahrten*)	0.053
Those using the man-engine (*Fahrkunst*)	0.186

It must be kept in mind that it is a terrific strain to climb up 1,200 feet and more on ladders, especially after eight or twelve hours of hard work. Among miners who had climbed that high in half an hour or an hour, Härting and

---

[3] George Rosen, *The History of Miners' Diseases* (New York, 1943), p. 167.
[4] *Ibid.*, p. 168.
[5] Meissner, "Die Gefahren des Bergwerksbetriebes für die Arbeiter und Schutzmassnahmen hiergegen" Theodor Weyl, ed., in *Handbuch der Hygiene* (Jena, 1897), VIII, p. 251.

Hesse found pulse rates of 168 and breathing rates of 42. Thus the cage constitutes not only a technical advance but also a hygienic one. Its use, however, requires many precautions and regulations. If a cage drops into the depths, all persons in it (usually ten to fifty) are killed. Therefore the hoisting ropes must be reliable and must be tested at short intervals.

The greater depth of mines makes ventilation more necessary and more difficult. The older methods of ventilation, as described by Agricola, were used until the 19th century. In 1760 James Spedding (the son) invented a new method of "air-coursing" that compelled the air to take a longer route through the mine. With the beginning of the 19th century, it became popular to divide the shafts into different parts for winding, pumping, and ventilating purposes. An attempt was made to use a "hydraulic box ventilator" or a trompe which sent a water spray into the shaft. Steam jets were installed for the same purpose, according to D. C. Jones (*Coal Mining*, 1941).

In 1807 John Buddle introduced the first mechanical ventilation in the Hebburn colliery and reported it in 1813. At about the same time Struve constructed a mechanical ventilator of the pump type in Wales. After 1840 more attention was paid to mechanically driven ventilators, such as those constructed by B. Biram (1842) and W. Brunton (1849).

It was not only the greater depth and therefore the higher temperature that made better ventilation necessary but also the change in the tools used and most of all the change in the products mined, the importance of which will be discussed later.

Dust. The danger of dust in ore mines was also increased by the introduction of new, power-driven tools. In discussing the dust problem we spoke about "miner's phthisis," which has been observed by English, German, and French physicians since the 16th century and which probably has always existed. This danger was increased by the introduction of drilling with power-driven tools.

The *Report* of the Royal Commission on Metalliferous Mines (1914) states in its review of previous inquiries: "Meanwhile [from 1895 to 1902] the prevalence of mortality from respiratory diseases among Cornish miners not only remained unabated but with the introduction of dust generating rock drills had even increased, and in 1902 a Committee . . . was appointed." The report of this committee (with Haldane and other physicians) said of Cornish miners that the stone dust "produces permanent injuries of the lungs gradually in the case of ordinary miners and rapidly in the case of machine-drill men." Later the report mentions the fact that "the machine-drill men, who are far more exposed to stone dust than any other class of miners, are also far more liable to lung disease."

The greater danger to which miners are exposed by machine than by the previously used hand drill can be explained by the following data. In a German quarry with compressed-air drilling, dust was generated to the amount of 451 particles per cubic centimeter as against 108 particles in a quarry with hand drilling. The South African Miner's Phthisis Prevention Committee also counted between 41 and 120 particles per cubic centimeter in hand drilling. Unfortunately the report of that committee does not give figures for dry machine drilling. However, the machine drills that used water supply (to decrease the amount of dust coming into the air) showed more dust, usually over 200 particles. Machine drilling is a much quicker process; many more holes are drilled in the same length of time, thereby further increasing the amount of dust in the air. Another very important source of dust is blasting, which is performed to a far greater extent now than in former times.

The rules of 1905, based on the recommendations of the Committee on Cornish Mines of the year 1902, require that in tin mines and ganister mines "while drilling holes by manual labor . . . water or other efficient means shall be used so as to prevent the escape of dust into the air. If rock drills driven by compressed air or other power are used, then efficient water jets, sprays or other effective means shall be used. . . . After blasting in any place no person shall return to that place until the air in such place has been cleared . . . or unless he uses an efficient respirator." The generation of dust by loading or by stone-breaking machines must be diminished by a thorough damping or wetting.

We have stressed the important contributions of the government, the commissions and the boards of South Africa to our knowledge of silicosis (see pages 202 ff., above). Important also is their effort to prevent this disease.

The South African regulation of 1903 required good ventilation. The regulation of 1906 prescribed the use of water jets or sprays for machine drilling, as did the English rules. The rule of 1911 added that in machine drilling the floor and sides of the working place must be kept sufficiently damp to a distance of at least ten feet to prevent dust from being raised. The rule of 1913 increased this distance to twenty-five feet but specified that in hand drilling only a swab must be provided and used around the drill at the collar of the hole. These rules were intensified during the following years by various means, especially by asking for abundant and constant water supply and by far-reaching directions on blasting and on the removal of broken rocks. Other requirements include regular examinations of the dust content of the air, an eight-hour day, sanitary precautions, and pre-employment examination of the workers. It may be added that better, less dust-producing types of drills also have been introduced.

The success of these measures is indicated by the following figures:

General average of the dust content of the air of South African gold mines in mg/ccm:

1913	5.4
1915	4.9
1927	1.2

Rate of silicosis per 1,000 workers:

1917–20	21.95
1926–29	21.1
1932–35	9.59
1935–38	8.00
1940–41	10.05

Mean duration of previous underground service in all new cases of silicosis:

1917–20	9 years, 6 months
1928–29	12 years, 7 months
1933–35	16 years, 8 months
1937–38	18 years, 7 months
1940–41	20 years, 2 months

The activity of the South African Government and the results achieved impressed the governments of other countries and incited them to activity. The Prussian Minister of Commerce arranged a competition for methods of rendering the dust of drilling innocuous. Results were published in the *Zeitschrift für das Berg-, Hütten- und Salinenwesen im preussischen Staate* in 1931. The following methods were found efficient: (1) wet drilling, which however does not seem to be advisable where softening of the rock and deterioration from climatic conditions must be avoided; (2) the binding of dust at the orifice of the borehole by means of foam, a method which may be efficient but needs to be improved; (3) the exhaustion of dust at the orifice of the borehole; and (4) respirators, which proved necessary because no method has been found which catches the drilling dust completely. But none of the masks tested was entirely satisfactory.

Explosions. The most important change in mining was in the material mined. Until recent centuries, mining was done primarily for ore (iron, zinc, lead, copper, and silver). Coal is reported to have been mined in the district of Aachen (Germany) since 1113, in the district of Essen (Germany) since 1317, and in England since the 13th century. In both countries, coal was produced in these earlier times by husbandmen, who worked in summer as farmers, in winter as coal miners. When authors describe hygienic conditions in mining before the 17th century, they are referring to ore mines. But since the 17th century, coal mining and the importance of coal have grown in England. According to J. E. Nef, 10,000 Scotsmen were working as coal miners and salters, as early as 1631. The annual output from 1551 to 1560 was 196,000 tons; from 1681 to 1690, 2,850,000 tons; from 1781 to 1790, 9,900,000 tons. In Germany the development began much later. In 1825 Prussia produced only 1,420,000 tons of coal, according to Bowden, who is quoted by Rosen. Today the production of coal is greater by far than that of ore, as the following table indicates.

PRODUCTION IN MILLION METRIC TONS

	Germany, 1938	England, 1938	U.S., 1943
Coal	186.4	230.6	587
Brown coal	196.0		
Iron ore	12.7	12.0	90
Lead and zinc	0.57		33

The greater depth of mines and particularly the prevalence of coal mining over ore mining gave rise to, and tremendously increased, another danger, the source of which must also be eliminated by ventilation. This is the methane ("firedamp") generated by the smoldering of organic substances without air (carbonization). The generation of this gas can occur in various mountainous formations, but it is more or less frequent and widespread in coal fields. There are coal districts in which this danger is almost unknown, but in others it is very great. So another gas, explosive methane, was added to the suffocating "blackdamp" (nitrogen with carbonic acid) and the poisonous gases such as hydrogen sulphide which had long been familiar to miners. Methane has no odor, is not poisonous, and is suffocating only if present in large amounts; but it is inflammable and explosive—which the other gases are not—if its content in the air is from 5 to 14 percent. Thus every open light, every spark, may be the cause of a great explosion.

The first ignition of firedamp, according to Rosen, was reported to have taken place in 1621 in a coal mine on the Tyne, when a man "burned in pit." The first actual mention of firedamp seems to have been made by Shirley in 1667 before the Royal Society. The next to mention it were Jessop (1675) and Beaumont (1681) in lectures delivered before the same society. Explosions occurred in England in 1675, 1677, and 1705. In 1708 a mine explosion in the county of Durham killed 69 persons. The deeper the mines were driven into the earth, the greater the number of explosions. Galloway reports that from 1835 to 1850, 463 explosions took place.

Good ventilation is the best way to avoid accumulation of methane and the consequent explosions. Even such ventilation cannot prevent the accumulation of methane in a niche or a sudden outburst of the gas from a pocket opened by mining. Therefore every possibility of igniting the methane must be avoided. Because the open oil lamps which were used in the mines ignite certain mixtures of air and methane, it was necessary to replace those lamps in mines where methane was presumed to develop by other lamps constructed in a special way.

First Clanny (1813), then Stephenson (1815) constructed lamps on the principle of insulating the flame from direct contact with the air by making it pass through water or oil. Davy, whose first lamps still followed the idea of Clanny, discovered (1816) that a wire gauze cylinder around the flame makes a lamp safe, and he constructed the lamp known by his name. It is worth mentioning that, although all these inventors were rewarded with medals and prizes by the Royal Society, none of them took out a patent, hoping that the lamp, not hindered by a patent, might be used everywhere for the benefit of miners.

Davy's lamp has been improved in several ways—by the use of benzol or mixtures of petroleum and alcohol, with internal ignition, by a magnetic lock which the miner himself cannot open, and by other changes. There are now various lamps in use which are modifications of the original Davy safety lamp: the Marsaut lamp in Great Britain, the Mueseler lamp in Belgium, and the Wolf lamp in Germany and the United States. Like the open lamps, all these lamps allow small amounts of methane to be recognized by changes in the flame (a pale blue cap covers the nonluminous flame). In this way the gas can be estimated within one-fourth of one percent. According to the English Coal Act of 1911, miners must leave their place of work as soon as there is as much as $2\frac{1}{2}$ percent of methane in the air.

A committee of the House of Commons wished to make the safety lamp compulsory in 1835, but the first results were not so good as expected. The operators thought that by the use of safety lamps they could save ventilation. The miners then complained of less ventilation, bad air, and higher temperatures. In 1839 the committee declared: (1) that the reliance on safety lamps alone in a gassy mine was a fatal mistake, and (2) that it was to be regretted that since the introduction of the safety lamp the ventilation of mines had become a matter of secondary importance. All these mistakes were gradually eliminated, and safety lamps are now compulsory in every European mine where there is danger of methane and, in a more restricted way, in some states of the United States.

All of the later English acts, regulations, and orders stressed the necessity of good ventilation. This became even more important because, with the greater depth of the mines, the temperature in them rose. An example will show the increase in ventilation currents.

At the beginning of the 19th century, John Buddle had asked for 18,000 cubic feet of air per minute in his mines. The figures given in the following table were also cubic feet of air per minute: [6]

Mine	1835	1850
Wallsend	5,000	121,360
Haswell	59,036	100,917
Hetton	98,300	190,000

These figures show us the improvement in ventilation methods up to 1850. But it is not possible to compare the data with the demands of today because we know nothing about the sizes of those mines. The calculations are made in another way today. In the Ruhr district mines, it was compulsory before World War II to have three cubic meters of air per man per minute.

As we have pointed out, the explosive concentration of methane in the air is between 5 and 14 percent. But as an accumulation of methane in certain parts of the mine is always to be expected, the ventilation must reduce the methane content of the air to a much lower level. The United States Mine Safety Board declared in a decision of 1926 that if in the return air from any group of workers the methane content exceeds 1.5 percent, such workers are considered to be in a dangerous condition. The Bureau of Mines holds that 0.5 percent of methane in the return air is the safety maximum.

[6] K. N. Moss, "Ventilation in Coal Mines," in Mining Association of Great Britain, *Historical Review of Coal Mining* (London, 1924), p. 139.

Safety Lamps and Nystagmus. The safety lamps had one bad effect. They made the illumination in mines, especially in coal mines, worse than it had been during the use of oil lamps. Neither the older English reports nor those of the Belgian commission made exact statements as to the luminous intensity of the various kinds of lamps. Marsaut, a French engineer, was the first to measure their candle power. Later Nieden submitted to the International Medical Congress in London (1880) data from which we quote the following figures on candle power:

	Freshly filled	*After 1½ hours' use*
Ordinary miner's lamp	1.42	0.732
Safety lamp (Westphalian)	0.418	0.322

A comparison of safety lamps of various types showed the following candle powers:

	Minimum	*Maximum*
Davy's lamp	0.13	0.26
Wolf's lamp	0.52	0.88
Marsaut's lamp	0.67	0.95

Other authors took as a unit the "foot-candle," that is, the illumination given by a standard candle at a distance of one foot. This measurement is more correct because it takes into account the great absorption of light by the coal surface.

In observations made before 1912, Llewellyn found that the illumination at the coal face in pits using old lamps was 0.09; in pits using safety lamps, 0.018. According to the investigations made by the same author for the Miners' Nystagmus Committee (*First Report*, 1922), the general illumination on the coal face in a candle pit was 0.145 foot-candles; in a safety-lamp pit, 0.009 to 0.02 foot-candles. Even electric lamps did not produce much better results (0.008 to 0.038). More recent examinations seem to show that improvements in the candle power of the safety lamp are possible. Wheeler and Woodhead of the British Safety in Mines Research Board (Paper No. 40, 1927) found that a large proportion of the Marsaut lamps in use gave no more than 0.5 or 0.6 candle power, but with modifications they were able to raise the Marsaut lamp to 2.5 candle power. Further progress has been made, and several other research studies on safety lamps followed later. One made in 1941 showed that safety lamps failed to maintain a reasonable light till the end of the shift, losing up to 40 percent or more of the original lighting power. This is true even of electric lamps.

Safety lamps are necessary only in mines in which the danger of methane exists. Where this is not the case, other lamps may be used or—better—electric light should be installed. Today the problem of lighting seems to have been solved to a great extent even in gassy mines by the improvements mentioned above and by portable electric lamps.

Too little light caused—and, where it prevails, still causes—an illness peculiar to miners, nystagmus, characterized by a trembling of the eyeballs, which if it reaches a high degree disables the worker.[7] According to Nieden (1894), the first cases were described in 1861 by Decondé, a Belgian physician. Peppmüller (Saxony) saw several cases in the years 1860 to 1863, as did Gillot (Sheffield), and Bronner and J. H. Bell (Yorkshire), but they did not publish their observations. The next publications were by Rode. Nieden was astonished to see so many cases among the coal miners of Westphalia. In 1894 he wrote an extensive and thorough book on this subject.[8] Thereafter the number of observations and studies increased. Among the publications were those of Dransart (1877–1913) in

[7] It may be added that nystagmus can be found as a congenital anomaly, as a symptom connected with other diseases of the eyes or the nervous system, and—in rare cases—can be acquired by bad illumination while doing fine work.
[8] A. Nieden, *Der Nystagmus der Bergleute* (Wiesbaden, 1894).

France, of Elworthy (1910–1920) and Llewellyn (1912 and later) in Britain, and of Rutten (1907–1919) in Belgium. There were also the many theoretical and practical works of Johann Ohm (1908 and later) in Germany, and finally the interesting reports of the British Miners' Nystagmus Committee (1922–1933).

In May, 1907, miners' nystagmus was added to the British schedule of compensable industrial diseases. The number of nystagmus cases reported in England increased. This was due in part to the wording of the regulation, which gave nystagmus a far-reaching definition, and in part to economic and other causes. Among men working underground, in 1908 there were 0.05 fresh cases per 100 workers; in 1920, 0.29, in the same year a total (fresh and old cases) of 7,028. In the German Ruhr district, 1.6 underground workers per 1,000 received invalidity pensions for nystagmus in 1908, 3.9 in 1930; in addition, 3.7 received sickness benefits for nystagmus in 1908, 5.7 in 1930 (Ohm).

There were many theories concerning the cause of nystagmus: intoxication (acute or chronic) by mine gases (Dransart); abnormal position in working, causing irritation of the labyrinth (Peschdo, T. H. Butler); overstrain of the eye muscles by looking upwards (Rutten, Snell, Stassen, Dransart); and bad illumination (Nieden, Nuel, Romiée, Court). But gradually it became clear that nystagmus spreads only among those coal miners who work for a long time with safety lamps and that it does not occur in mines with good illumination. J. S. Haldane says in the first report of the Miners' Nystagmus Committee (1922): "As has been shown above, nystagmus is a disease connected with the use of safety lamps. . . . We may therefore conclude with complete certainty that the abnormal constituents in ordinary mine air have nothing to do with the production of miners' nystagmus." And the third report says: "A principal finding of the earlier committee was that the chief symptom of the disease . . . was caused by an insufficiency of the light reaching the eye of the miner while at work and that of all measures of prevention the most important was to secure for the miner at work an adequate illumination. The present committee reaffirm in the strongest terms both this scientific conclusion and its practical corollary." The theory that nystagmus is due to the overstraining of the eye muscles by the permanent "constrained position of the eyes by which chronic weariness is induced in the elevator muscles of the eye" (Snell) is refuted by the fact that not only hewers but all classes of underground workers are affected, even supervisors. So it is clear that the disease which is caused by the safety lamp will disappear when the latter is replaced by better lighting equipment, which of course must be equally safe.

The problem seems to find its best solution in electric lamps, especially electric cap lamps. They must be so constructed as to make it impossible for them to ignite any firedamp or dust in the surrounding area. Portable electric lamps for miners were constructed in England in the year 1889 by John Davis and Son in Derby. But it was not until 1910 or 1911 that their use became more general. In 1911 there were 4,298 electric miners' lamps in use in English mines. In 1922 there were 294,593, which is roughly a third of all miners' lamps then in use in England. On the Continent only a small number of portable electric lamps were used at that time. The chief hindrance was their great weight.

In the United States the first electric hand lamps were built in 1912. In that year the United States Bureau of Mines marked such a lamp as permissible, and in 1915 it approved an electric cap lamp. Cap lamps are used very rarely on the Continent, but are used regularly in the United States. This is due to the different bodily position of the miners working on the smaller coal seams of Europe and on the broader coal seams in the United States. Now the Bureau of Mines calls permissible those electric cap lamps which meet various safety conditions and also satisfy the following requirements in lighting: they must give a beam which has a spread of at least 130°, and the surface covered by the beam must not show sharply contrasting areas of bright and faint illumination; the maximum candle power must not be greater than five times the average or mean candle power of the beam, which, based upon readings at the designed voltage of the bulb, must not be less than one candle power.

Safety lamps, as was pointed out above, and open-flame lamps allow the presence of gases to be recognized by changes in the flame. If the safety lamp is replaced by an electric lamp, care must be taken that "indicators" of gases are provided. A simple method is in use in most countries and has been recommended by the United States Bureau of Mines. In mines where firedamp or blackdamp may be present and electric lamps are used, good safety lamps are given to at least several of the experienced or especially trained employees, who must observe the flames carefully.

In former times sparrows, mice, and even dogs were used to indicate by their loss of consciousness or their death the presence of dangerous gases. The oldest artificial indicator seems to have been the "flint-and-steel mill" constructed by an Englishman, Carlyle Spedding, in 1750. It produced sparks when the wheel was rotated, and the kind and size of the sparks indicated the type and amount of mine gases, if any were present. In some books the mill is spoken of as a lamp; but inasmuch as it produced sparks only when the wheel was rotated by hand, its use as a mine lamp was not possible. The use of the safety lamp as an indicator followed later. Today several indicators are in use: in the United States and England the Hoolamite detector for CO, the M.S.A. combustible gas indicator, the M.S.A. hydrogen sulphide detector, in Germany the Degea "Kohlenoxyd Anzeiger" (CO indicator), the "Gasspürgerät" (Gas-finding instrument) of Draeger, and others previously mentioned (Chapter 5).

Control of Explosions. Mining had always been a work of pickaxes. In 1627 Casper Weindl introduced the use of gunpowder for blasting in mines. But it was found that in the deep coal mines gunpowder was dangerous because of firedamp and coal dust. The final report of the British Royal Commission of 1879, published in 1886, recommended the use of high-power explosives (dynamite) instead of powder, as did the Prussian Schlagwetter (Mine Disaster) Commission. But dynamite too proved dangerous, and so "safety explosives" (*Wettersprengstoffe*) were made—the first in Germany. In 1896 England prohibited the use of those explosives which were not on a "permitted list." Since 1923 only explosives approved by the Minister of Commerce have been allowed in Prussia. The regulations concerning explosives are similar in most of the European countries. In the United States the Bureau of Mines also published a list of "permissible explosives" and "permissible blasting units"; its advice, however, is not compulsory.

The terrible extent of explosions with all their frightfulness was not due to firedamp alone, nor were all explosions caused by firedamp.

Attention was first called to coal dust as the accelerating force of explosions initiated by methane in the mine disaster that occurred in Wallsend (Newcastle upon Tyne) in 1803. Later Faraday and Lyell (1835) and du Souich (1855) called attention to the same agent. In the following years great disasters, especially in England, hastened investigation of the role played by coal dust in mine disasters. The British Royal Commission on Accidents in Mines in its report (1866), the Englishman Galloway, the Prussian

Mine Disaster Commission (1880–1886), which used an experimental gallery, an American commission, and finally two Americans, W. N. and J. B. Atkinson, all came to the conclusion that coal dust alone or coal dust mixed with a small amount of methane causes explosions and gives them their terrible effectiveness.[9] It may be remembered that the amount of coal dust in mines has been increased and is still being increased with the introduction of machines in coal mining.

The British Coal Mines Act of 1887 prescribed that, before firing, the surrounding area must be wetted for a distance of sixty feet. Similar rules were published in Prussia. Later English and French rules ordered, in addition to this wetting, a regular, complete cleaning of the big pathways of coal dust. The German rules required a thorough wetting of all galleries.

The explosion at Courrières, France, which killed 1,230 miners in 1906 brought accelerated efforts to find methods of preventing coal dust explosions. This mine had always been supposed to be free from methane, and investigation did indeed show that it was a coal dust explosion initiated by faulty blasting. In the next years other great disasters followed. In 1907 an explosion in the Saar district killed 150 men. One in West Virginia killed 358. In 1908 a great disaster in the Radbod mine (Westphalia) killed 348 miners. In a few countries, experimental galleries had been equipped earlier—one in Prussia (1884–1886) and one in England (1897). But these disasters gave impetus to the installation of experimental galleries which, created by private organizations, were soon taken over and enlarged by the governments. The first installations of this kind were made in France in 1907, in England in 1906 and 1908, in

---

[9] See K. Hatzfeld, "Die Entwicklung der Massnahmen zur Kohlenstaubbekämpfung," *Berg- und Hüttenmännische Zeitschrift "Glückauf,"* XLIV (1925).

the United States in 1908 and 1910, and in Prussia in 1909 and 1911.

In the Altofts mine in Great Britain, a disaster occurred in 1887. The director, Garforth, and other experts found that in those galleries where there was, in addition to coal dust, a large amount of rock dust in consequence of the crumbling of the gangue no explosion had taken place. From that observation the idea originated that other dust mixed with coal dust diminishes the danger of explosion. In 1906 Garforth installed a small experimental gallery, later enlarged, where he worked together with the chemist Wheeler. From 1911 to 1915 the latter continued the work in an experimental gallery of the government for the Explosions in Mines Committee. The final results of these investigations were the British rules of July 30, 1920. They prescribed that rock-dusting be done in such a way that the dust that is whirled up should contain at least fifty percent of noninflammable material. The order of November 20, 1924, intensified this requirement.

In France, Taffanel—after experiments made from 1909 to 1913—recommended dust barriers, that is, an accumulation of rock dust on planks so loosely placed in the drifts that an explosion overthrows the plank and so disperses the dust at the critical moment. The French regulations of 1911 and 1912 ordered, for use in less dangerous mines, the cleaning and wetting or dusting of the area adjacent to the blasting place. Wetting also was required in dangerous mines, but the regulations ordered particularly the use of rock dust by "schistification" (dusting) and first of all by dust barriers.

In Germany, Beyling began research with rock dust in an experimental gallery in 1911 and published his results in 1919. In 1913 and 1914 the Prussian Minister of Commerce sent a commission to England and France to study their experience with rock-dusting and other methods of preventing explosions. The results of these studies were published in 1918 by Hatzfeld. The commission recommended the regulation of the use of explosives and practical experiments with dusting. In 1919 a commission of officials, mine managers, and mine workers was appointed in the Ruhr district to help the mines introduce dusting and to observe its results. Upon the recommendation of the commission, and because dusting had proved very effective in several explosions, the further use of rock dust for barriers quickly spread in the mines. It is now prescribed that all mines having seams with dangerous coal dust must be made safe with rock dust used in accordance with rules both for dusting and for barriers.

Investigations made by the U. S. Bureau of Mines in the Pittsburgh Experimental Station should also be mentioned; they are reported by G. S. Rice and others in the Bureau of Mines Bulletin, Nos. 56, 167, 268, and 269, along with instructions laid down by the Bureau.

In all countries, in order to avoid the danger of silicosis, fine rock dust with a very small quartz content must be used for dusting. In Prussia strict regulations govern the testing and approval of such dust. The result of all these efforts is summarized in Table 1.

Mine disasters are most dramatic and therefore attract the attention of the people and of the legislative bodies. But accidents caused by explosions are not nearly so numerous as those caused by other conditions in mines. We have taken into account here fatal accidents only, because the number of other accidents reported is influenced by the regulations of the various countries determining what types of accidents must be reported and even more by the practice of reporting. In 1936 among 1,000 underground workers in Prussia there were 0.84 fatalities caused by falling rocks or coal, 0.38 by gases and coal dust fires or explosions. In the United States in 1942, among 1,000 men there were 0.218 fatal accidents resulting from the explosion of gas or dust, 1,004 from the fall of roof. In 1936 the fatal accidents caused by falls of roof amounted in Prussia to 39.6 percent, in Great Britain to 48.1 percent, in the United States to 55.6 percent of all fatal accidents. Other fatal accidents are caused by tools and machines, by electricity, by explosives and sparking plugs, and by elevators and conveyors.

TABLE 1

A: GREAT BRITAIN

COLLIERY DISASTERS CAUSED BY EXPLOSIONS [a]

Period	Principal Disasters [b]	Deaths
1851–1860	31	1,271
1861–1870	33	1,444
1871–1880	35	2,014
1881–1890	26	1,292
1891–1900	10	733
1901–1910	13	1,059
1911–1920	5	707
1921–1930	7	170
1931–1938	13	606

[a] From the *Report of the Secretary for Mines (1925)*, p. 154, and later reports.
[b] Accidents involving the loss of 10 or more lives.

B: PRUSSIA [c]

Period	Average Annual Explosions	Average Annual Deaths
1891–1900	70.5	72.7
1901–1905	30.8	21.9
1906–1910	30.6	116.8
1911–1915	23.8	58.6
1916–1920	37.2 [d]	58.6
1921–1925	14.8	46.6
1926–1929	9.0	19.0
1931–1937	5.1	21.7

[c] From *Das Grubensicherheitswesen in Preussen im Jahre 1929*. Supplement on "Zeitschrift für das Berg-, Hütten- und Salinenwesen im preussischen Staate 1930," B 459, and later reports.
[d] Among them, one explosion with 107 fatalities.

The number of all these accidents can be and has been

diminished by regulations and installations, but much depends on the caution and the training of engineers and workers. To carry out regulations and to observe all the necessary precautions, not only well-instructed mine management is necessary—and the British and German rules take care of that—but also well-trained workers. In the Ruhr district in Prussia, only such persons are allowed to work as hewers who, after three years of mine work and after training and an examination, have received a certificate from the mining authority. In Great Britain, the Mine Department together with a committee of the Mining Association (mineowners) try to provide for the technical education of miners.

RESCUE WORK

In spite of all precautions, disasters still occur. Consequently a well-organized and well-equipped emergency rescue squad must be provided. We are told by medieval writers that when one miner had an accident, all his comrades tried to rescue him, even at the risk of their own lives. John Buddle describes in 1835 what he called the barbarous practice of moving injured and unconscious men in carts galloping over the roughest ground, which may have killed many. In 1839 a British committee gave instructions for resuscitation. In modern times, with more complicated conditions, spontaneous and improvised rescue work is not sufficient. By the British Act of August 3, 1910, the Secretary of State obtained the right to issue rules on the installation of rescue work and ambulance services as well as the training of men. The regulations of 1928, 1930, and 1935 gave detailed orders for central stations, each to have a radius of action of fifteen miles—in exceptional cases twenty miles—and a well-trained superintendent and trained men. Every mine must be connected with such a station and must have a trained rescue team with apparatus

MINES AND MINERS 253

of its own. In 1938 there were sixteen such central rescue stations.

The organization in Prussia is similar. Every mine district has a central station for rescue work and every mine its trained rescue workers with necessary apparatus. According to the latest reports (1937), the Prussian rescue stations and mines had in stock 4,096 breathing apparatuses, among them 3,383 of the self-breathing type, 85 percent of them Draeger apparatuses. No type of gas mask is permitted in any rescue work either in Great Britain or in Germany.

In the United States the training of men in rescue work was first performed in the Pennsylvania anthracite district in 1899. In 1910 the Bureau of Mines initiated the training of men. Now it recommends that every large mine have a central safety committee, a workmen's safety committee, and safety engineers. Workers are to be trained in rescue work, and rescue apparatus should be ready, including self-contained oxygen breathing apparatuses. A special gas mask is also approved. Very good booklets for such training have been published by the Bureau of Mines.

Attempts to equip rescue workers were made early. In 1825 Roberts recommended goggles and a sponge soaked with lime water.[10] Henry de la Beche and Lyon Playfair found in 1846 that after an explosion the oxygen content of the air is sufficient for breathing (!) and recommended that a bag filled with a mixture of sodium sulphate and lime be bound over the mouth. The development of self-breathing apparatus which first made it possible for rescue workers to enter rooms filled with irrespirable gases has been described (see Chapter 5).

Such breathing equipment must be stored in an easily accessible space near the entrance to the mines and must be

---

[10] J. A. S. Ritson, "The History of the Development of Mine Rescue Work in Collieries of this Country," in *Historical Review of Coal Mining* (London, 1924), p. 241.

taken care of so that it is constantly ready for use. Other equipment for rescue work also must be kept in good condition, ready for use.

ANKYLOSTOMIASIS

Among the dangers to which miners are exposed there is finally a disease which some decades ago played an important role in European tunnels and mines. Ankylostomiasis or hookworm disease is extremely prevalent in tropical and subtropical countries between 35° latitudes both north and south of the equator. There the eggs and larvae of the worm, evacuated from the body of an infected person by way of the stool, survive in the warm, humid soil and get into another person's body. They cause a serious, even fatal anemia. The struggle against this disease is one of the most important tasks of public health in these countries. Such projects have been much promoted by the Rockefeller Foundation.

The worm can spread only in mines where the temperature is 25° to 28° C. This is the case only in deep mines, in Europe below 2,000 feet. Humidity also increases the danger. It seems that in 1789 there was an outbreak of the disease among miners in Schemnitz (Hungary); Hoffinger reported 1,200 cases. Other epidemics followed: in 1820 in Anzin, Fresnes, and Vieux-Condé, described by Noel Hallé, and in the same year in other coal mines of northern France. Smaller epidemics had occurred in Bohemia and Hungary by 1873. A terrible outbreak of the disease occurred during the construction of the Gotthard Tunnel. Hundreds of workers are said to have died, thousands fallen ill of ankylostomiasis between 1877 and 1880. From this place the epidemic spread over all the mines of Europe. The coal mines in the German Aachen and Ruhr districts, those near St. Etienne and Anzin in France, the Belgian mines of Liége, Mons, and Charleroi, the Cornish tin mines in England,

and several Bohemian and Hungarian mines were all infected. In these places the epidemic reached its peak between 1902 and 1905.

The disease was combatted in the following ways:
1) By building numerous sanitary accommodations in the mines and making their use compulsory;
2) By keeping all infected persons away from the mines; and
3) By treating infected persons until they were completely free from the worm.

In the German mines (Ruhr district), in compliance with a regulation of the mine authorities, 17,000 sanitary appliances were installed, and the stools of all the miners were examined repeatedly. Approximately 7,000,000 microscopic examinations of stool were made, and about 50,000 treatments provided. In the first examination of 1902–1903, 14,548 miners were found to be carriers of ankylostoma; in November, 1904, only 3,288; in March, 1909, only 749; and from 1911 to 1926 only a few persons proved to be infected without being anemic. By a decree of the Prussian police of 1900, confirmed by a law of 1903, the common plunge baths, where they existed, had to be replaced by separate shower baths at the mines. The disappearance of ankylostomiasis in Germany, in these deep and humid mines of the Ruhr district, is a phenomenal triumph of well-planned epidemiological control. Hayo Bruns, from whose work—"Durch Eingeweidewürmer bedingte Berufskrankheiten" (Occupational Diseases Caused by Intestinal Worms), *Handbuch der Sozialen Hygiene* Vol. II (Berlin 1926)—the data above are taken, was one of the principal leaders of the program.

RESULTS OF MINE HYGIENE

What, then, has been the result of all these investigations, laws, and regulations? Haldane composed tables for

Great Britain which I have completed for recent years as far as figures were available.[11] Table 2 shows how much the mortality of British males has decreased since the middle of the last century. The decrease in the mortality of coal miners is even greater; the death rate is now only slightly higher than that of all males, although 80 years ago it was 36 to 45 percent higher than that of the population for all age groups. It is amazing that the mortality of coal miners who work below ground is less than that of the men who work above ground. This may be due to the selection of stronger men for the actual mining.

A decrease in the mortality from accidents is to be seen in every country. In Great Britain (Tables 3 and 4) and the United States (Table 6), the earliest available reports show figures as high as 0.5 and 0.6 percent. In Great Britain we see a constantly decreasing rate of fatal accidents in coal mines as well as in metalliferous mines (Table 4). More recent data than those shown in the tables are not available. In Prussia (Table 5) with its strict governmental supervision even some time ago, the first available figures show lower rates (1.68 per 1,000 workers in all mines); after the change in mine laws (1865), we note a moderate increase, followed shortly by a new decline that was slightly interrupted during the period from 1911 to 1920.

While the statistics of the European countries are based on laws requiring reports to be sent to the central authority, the statistics of the United States Bureau of Mines are based on the reports of the inspectors of the single states. These—as Fay wrote in 1914—are "with few exceptions complete" and have been carefully revised. So we may accept them as being reliable from the beginning of this century. In the coal mines of the United States we find, after the first high figures, a decrease in the rate of fatal

---

[11] J. S. Haldane, "Health and Safety in British Coal Mines," Chapter XVIII in *Historical Review of Coal Mining* (London, 1924), p. 267.

accidents (Table 6), but towards the end of the century there was again an increase. These figures (column 2) seem especially high when we remember that in these years the American miner had fewer working days and fewer working hours per year than the European coal miner. A marked decrease began with the five-year period 1931 to 1935, falling below that of the end of the previous century.[12]

TABLE 2

GREAT BRITAIN: ANNUAL DEATH RATES PER 1,000 WORKERS, BY AGE GROUPS

	15-20	20-25	25-35	35-40	45-55	55-65
1849–1852						
All males	8.1 [a]	..	10.1	12.7	18.9	31.8
Coal miners	14.5 [a]	..	14.5	17.2	26.3	44.0
1900–1902						
All males	2.46	4.50	6.29	10.87	18.72	35.56
Coal miners	3.21	4.51	5.08	7.97	15.19	38.02
1930–1932						
All males	2.69	3.28	3.46	5.69	11.14	23.55
Coal miners	3.42	3.81	4.17	6.45	11.46	23.73
Coal miners						
Below ground		3.68	4.05	6.33	11.40	23.43
Above ground		4.81	5.46	7.60	11.94	25.56

[a] In age group 15–25.

[12] It should be stressed that the calculation of accidents per 1,000 employed persons or per 1,000 workers is not an exact one. Their number may change during the year and the estimate by the management of the average number may not always be correct. For this reason, in Germany figures are given for 1,000 *Vollarbeiter*, meaning for 1,000 workers each of whom works 300 days annually, therefore 300,000 working days. If we look at Table 8 on Germany, we find in some years and industries little or no difference in the calculation for workers and *Vollarbeiter*, which indicates a very regular occupation. But in the United States the differences are very great, in consequence of the peculiar circumstances in the American coal mines in earlier times. The number of yearly workdays of the miners in these times was highly variable. During the period 1913 to 1923 it fluctuated between 129 (1921) and 249 (1918); in Illinois in 1923, 1924, and 1925 miners worked only 146, 139, and 139 days. So it was also with the work hours, which oscillated between 1,943 a year (1916 to 1920) and 1,321 a year (1931 to 1935). Therefore Fay (1914) proposed a calculation on a basis of 200 work days a year. Now the average occupation of a worker is about 280 days a year. But not even a calculation for *Vollarbeiter* is correct, especially if we consider longer periods, because

TABLE 3

GREAT BRITAIN: ANNUAL DEATH RATES BY ACCIDENTS PER 1,000 COAL MINERS, BY AGE GROUPS

Period	15-20	20-25	25-35	35-40	45-55	55-65
1849-1853	5.7 [a]	..	5.3	6.2	6.9	5.9
1900-1902	1.1	1.0	1.1	1.4	1.7	2.0
1930-1932	..	..	0.99	0.14	1.23	1.31

[a] In age group 15-25.

TABLE 4

GREAT BRITAIN: FATAL ACCIDENTS PER 1,000 MINERS

	COAL MINERS		METALLIFEROUS MINERS
Period	All Causes	Explosions	All Causes
1851-1855	5.15	..	..
1873-1882	2.24	0.65	1.62
1883-1892	1.81	0.32	1.44
1893-1902	1.39	0.18	1.31
1903-1912	1.33	0.17	1.30
1913-1922	1.15	0.10	1.25
1923-1932	1.05	0.06	1.10
1936-1938	1.05	0.12	1.17

TABLE 5

PRUSSIA: FATAL MINE ACCIDENTS

PER 1,000 WORKERS			PER MILLION WORK HOURS			
Period	Coal Mines	All Mines	Year	Rate	Year	Rate
1840-1850	..	1.68	1911	0.19	1929	0.82
1861-1866	2.66	2.17	1912	0.93	1930	1.09
1867-1880	2.94	2.47	1913	0.90	1931	0.80
1881-1890	2.47	2.46			1932	0.64
1891-1900	2.11	2.19	1925	0.97	1933	0.72
1901-1910	2.48	1.97	1926	0.84	1934	0.64
1911-1920	2.80	2.64	1927	0.85	1935	0.58
1921-1925	2.07	1.92	1928	0.78	1936	0.58
1926-1930	2.31	..				
1936	1.50	..				
1937	1.50	1.39				

the daily work hours change—and what is important is the time of exposure. The present basis of calculation of the Bureau of Mines is correct, computing accidents on a period of a million work hours. But the difficulty is that the figures calculated in different ways—worker, *Vollarbeiter*, and per million work hours—are not easily and correctly comparable.

The reasons for using the statistics of fatal accidents only are explained in Chapter XII.

TABLE 6
UNITED STATES: FATAL COAL MINE ACCIDENTS

Period	Per 1,000 Workers	Per 1,000 300-Day Workers	Per Million Man Hours
1870	5.93	..	..
1871–1875	4.30	..	..
1876–1880	2.72	..	..
1881–1885	2.85	..	..
1886–1890	2.39	..	..
1891–1895	2.91	4.38	..
1896–1900	2.95	4.50	..
1901–1905	3.45	4.95	..
1906–1910	3.94	5.48	..
1911–1915	3.40	4.65	1.8
1916–1920	3.18	4.03	1.64
1921–1925	2.73	4.58	1.89
1926–1930	3.19	4.61	1.90
1931–1935	2.24	3.93	1.69
1936–1940	2.27	3.64	1.72
1941	2.32	3.26	1.54
1942	2.77	3.42	1.62

Great differences exist in the United States between the different states. A compilation by the Bureau of Mines (Bulletin No. 456) shows that if, for the years 1937 to 1941, the fatality rate per million work hours is 100 for underground work in coal mines in the United States as a whole, then the fatality rate by states is, for Texas, 25; North Dakota, 44; Pennsylvania, which produces one-fifth of all the coal, 57; but for Illinois, 118; Indiana, 239; and—among the states and territories with smaller production—Arkansas, 237, and Alaska, 891.

A comparison of the rates of various countries is difficult to make because of the different methods of calculation used. If we take the calculation "for 1,000 workers," the figures for fatal accidents in the United States are constantly more than double those of Great Britain, despite the higher number of annual working days in the latter, especially in earlier years. The American figures also have always been higher than those in Prussia. We obtain the most accurate picture by comparing the figures of fatal accidents per million man-hours. In the period from 1931 to 1935, for example, the rate of fatal accidents per million working hours in coal mines in the United States was 1.69; in Prussian mines, 0.68. This Prussian rate is for all mines. Figures on Prussian coal mines per million working hours are not obtainable, but a comparison of the figures for coal mines and for all Prussian mines for 1,000 shifts show that the rate for coal mines is about 10 percent higher than the average for all mines. Thus the rate in Prussian coal mines would be 0.76, as compared with 1.69 in the United States. In other words, the accident death rate in the United States was about 220 percent of that in Prussia —or in the American coal mines approximately 950 more accidental deaths occurred each year than would have occurred here if the same conditions prevailed as in Prussia. It may be mentioned that since that time (1931 to 1935) until the last available figures, the rate of fatal accidents in the United States has declined only slightly.

Any such comparison between countries must take into account the technical, economic, and legal differences. According to G. S. Rice and J. Hartmann, the average hoisting depth of the shaft in the United States was less than 500 feet; in Great Britain, 1,000 feet; in France and Germany, 1,500 feet; in Belgium, 2,000 feet.[13] In the United States, 48.2 percent of the coal mined in 1942 came out of 316 mines, each with a production of more than 500,000 tons. There are 1,424 mines in this country which produce 50,000 to 500,000 tons each, in all 44.9 percent of the total production; 1,492 mines in which 10,000 to 20,000 tons are produced; and finally 3,740 mines with an annual production of less than 10,000 tons, accounting for 2.2 percent of the total production of the United States. In Great Britain in 1936 there were 2,080 coal mines, belonging to

[13] *Coal Mining in Europe* (U.S. Bureau of Mines Bulletin No. 414), 1939.

1,000 separate concerns (in 1913 there were 3,267 mines). In 1937, 77 percent of the total output was produced by 149 concerns. In Germany there were 262 coal mines in operation in 1935, with an average production of 546,000 metric tons per mine. On the whole, the German coal mines are larger and deeper, especially as compared with those of the United States, where there are many smaller pits.

But the great coal mines in this country are more mechanized than elsewhere, and this has increased greatly in the last two decades. In 1927, 5 percent of the mined coal was mechanically loaded; in 1944, 50 percent. In 1917, 55 percent was cut by machine; in 1942, 89.7 percent. As to the effect of mechanization on accidents, opinions were divided. G. S. Rice expected an increase of accidents, Charleston a decrease in the number of serious ones. The British Royal Commission on Safety in Coal Mines in its report of 1938 says that in the beginning there was a period in which the unfamiliar conditions resulted in risks, adding, "We believe however that there must ultimately be a gain in safety." Concerning fatal accidents, A. H. Fay presents an interesting table in the Bureau of Mines Bulletin No. 115 (1916). He divides the states into those in which (1) less than 20 percent of coal is mined by machine, (2) 20 to 39 percent, (3) 40 to 59 percent, and (4) 60 percent or more. In the first group, representing the smallest percentage of mechanized mines, the accident death rate per 1,000 workers was 4.09; in the second group, 3.67; in the third group, 3.21; and in the fourth group—in states with the highest percentage of mechanically mined coal—the rate was 2.82. It is worth noting that in group one, 1.11 were killed in exceptional accidents; in group four, only 0.20; but the death rate from common accidents was 2.98 in group one and 2.62 in group four. These figures from an earlier period in which mechanical work was not so highly developed, especially as regards precautionary measures and the training of workers, show clearly that the number of fatal accidents is not increased but rather decreased by mechanization. Pointing in the same direction—but not so clearly because we do not know the distribution of workers in different kinds of work—is the fact that in 1942 the rate of workers killed was 1.616 per million work hours; but only 0.069 were killed in work with mining machines of every kind, whereas 1.004 were killed by fall of roof and coal.

Rice and Hartmann point out that in general the ventilating methods in European coal mines are excellent—better on the average than those in the United States—but they are not better than those in the best equipped mines in this country. In Europe, safety practices are officially prescribed. The governmental safety regulations are carefully drawn up and are enforced more strictly than those in the United States. The qualifications of the mine inspectors are much higher in Europe, and the inspectors are required to pass difficult examinations. The manager and his assistants must be well-trained mining engineers. In short, governmental supervision and influence are much greater in Europe than in the United States.

Although the natural conditions (shallower mines) as well as the technical conditions seem to be more favorable in this country, nevertheless in all these years the accident mortality rate has been much higher here than in Europe, with its strict governmental rules and strict governmental supervision of the mines. Rice and Hartmann write that "these factors are probably important in bringing about the much lower fatality rates as shown by the accident statistics." It is surely correct to assume that the lower accident rates in Europe are a direct consequence of the stronger governmental control. Intelligently written, detailed regulations, their enforcement, and strong supervision by the government seem to be the most efficient factors in diminishing mine accidents.

Circumstances have changed in this country during very recent years. In 1941 the Bureau of Mines, with its outstanding experts, obtained the right of mine inspection and has undertaken many inspections since that time. In July, 1946, a mine safety code for bituminous coal mines, drawn up largely by these experts, was published. It is to be hoped that the influence and the power of the Bureau of Mines will increase in the next years and that it will be able to enforce safety in mines.

12

The Effects of Industrial Hygiene

IT IS A WELL-KNOWN FACT that the standard of living as well as the health and the life expectancy of working people have improved considerably in the last 150 years. The mortality rates of the English male population from 1849 to 1852 and from 1930 to 1932 were shown in Table 2. These mortality rates decreased enormously in the intervening years—to less than half in the age group from 15 to 45. And these figures are determined in large part by the working class, which constitutes the overwhelming majority of the population. To achieve this result, all improvements in public health, including housing and nutrition, wages and working hours, personal hygiene and morals, worked in conjunction with factory hygiene. Therefore these figures do not indicate the improvements brought about by industrial hygiene alone, and we have no way of showing the effect of the latter statistically.

These difficulties are set forth in an excellent article by Greenwood. He points out that "the general rate of mortality of England and Wales remained until the last thirty years of the 19th century no better and indeed probably worse than in the first fifteen years—in spite of the labor laws. . . . Had our ancestors wholly neglected the evils of life *in* factories . . . but rehoused the people, I think the rate of mortality would have begun to fall a century sooner."[1] He also stresses the importance of personal hygiene and morals.

Although we are not able to recognize the effect of factory hygiene apart from other influences on the health of the workers, figures that show us the comparative mortality

[1] M. Greenwood, "The Evolution of an Industrial Society," *British Journal of Industrial Medicine*, Vol. I (1944).

THE EFFECTS OF INDUSTRIAL HYGIENE 265

rates of the upper class and the working class also indicate how much has been achieved by the economic rise of the latter group, by public health work, and by factory hygiene altogether—and how much must still be done.

Useful statistics on this are available in Great Britain and in the United States. Since 1910 the British Registrar-General's *Decennial Supplement* has carried not only reports on occupational mortality, which have been published since 1864, but also on mortality according to social class. Because this scheme of tabulation was changed, only the statistics of 1921–1923 and 1930–1932 are useful to us.

If we designate 100 as the standardized mortality rate of men between the ages of 20 and 65 years in Class I ("upper and middle class, the professional and generally well-to-do section of the population"), then we arrive at the following figures:

	1921–1923	1930–1932
Class I (upper and middle class)	100	100
Class II (intermediate)	113.5	108
Class III (skilled workers)	115	112
Class IV (intermediate)	121	116.5
Class V (unskilled workers)	148	128

It must be kept in mind also that the mortality rate of men between 20 and 65 was 855 per 100,000 in the period from 1930 to 1932, that is, 90 percent of the 1921–1923 rate. The death rate of Class I remained unchanged, while that of Class V dropped to 82 percent of the 1921–1923 rate. To summarize, the mortality rate in general decreased during this decade, but among unskilled workers to a greater degree than among other classes. The rate among unskilled workers in 1921–1923 was 48 percent higher than that of the well-to-do class in England; in 1930–1932 it was 28 percent higher.

If we calculate the percentage of deaths for 1921–1923 and for 1930–1932 in the different age groups of Classes

IV and V, we get the following figures (based on a death rate of 100 for every age group in Class I):

PERCENTAGE OF DEATHS BY AGE GROUPS

	20–24	25–34	35–44	45–54	55–64
1921–1923					
Class IV	155	161	138	119	111
Class V	172	190	182	153	136
1930–1932					
Class IV	99.3	125	125	128.8	105.2
Class V	100	130	136	132.3	114

For the United States there are no official statistics, but we have data for 1923 from the Metropolitan Life Insurance Company.[2] The death rates in 1923 of the "regular ordinary department" (total males) and the "industrial department" (white males) are given in the following table:

DEATH RATE PER 100,000 BY AGE GROUPS

	20–24	25–34	35–44	45–54	55–64
Industrial Department (white males)	400.6	556.1	946.6	1,725.4	3,385.3
Ordinary Department (total males)	255.2	268.4	422.0	790.1	1,868.0
Percent, Industrial of Ordinary	157.0	207.2	224.3	218.4	181.2

The "ordinary department" of the Metropolitan data corresponds approximately with Class I of the British statistics. The "industrial department" (white males) can scarcely be said to correspond with Class V, unskilled workers, nor with Class III, skilled workers. It does correspond roughly, however, with Class IV, the intermediate group between the two. A comparison with the British figures for 1921–1923 shows a great difference unfavorable to the American working class. In England in the age group 25 to 34 the mortality is 161 percent and in the United States 207.2 percent of that of the well-to-do class, and in

[2] L. J. Dublin and R. J. Vane, *Causes of Death by Occupation* (U.S. Bureau of Labor Statistics Bulletin No. 507, 1930), p. 7.

the years 35 to 44 the differences are 138 percent and 224.3 percent.

For later years the figures given by Whitney (*Death Rates by Occupation*, National Tuberculosis Association, 1934) are useful. They are taken from the records of only ten states for the year 1930. The standardized death rate for all gainfully occupied males is 870 per 100,000. There are several groups, from which we select those interesting to us.

	Standardized Death Rates per 100,000	Comparative Death Rates on Basis of 100 for Professional Men	Comparative Death Rates on Basis of 100 for Proprietors, etc.
Professional men	700	100	..
Proprietors, managers, officials	738	105	100
Skilled workers and foremen	812	116	111
Semiskilled workers	986	141	134
Unskilled workers	1,310	187	178

The age groups reported—25 to 44 and 45 to 64—are large and not very useful for purposes of comparison; the rates are not standardized. Therefore I quote here the standardized rates of the group of 15 to 64 years. I also believe that it would not be in conformity with the British study to take as the basis of 100 the professional men (lawyers, judges, physicians, technical engineers, teachers, and so forth) but would be better to use the second group (proprietors, managers, and officials). We see that the difference unfavorable to the semiskilled group, which probably corresponds with Class IV in Britain, is greater in the United States than in England. Even greater is the increase in the mortality of the unskilled workers (78 percent), partly because of the number of colored people included in this great mass, which in Europe would be called the lower proletariat.

As a whole these figures show that the health position of

the American workers compares less favorably with that of other classes than does the British. But, if the figures of Whitney are comparable with those of Dublin and Vane (Metropolitan), even that would indicate great progress in the third decade of this century. It is regrettable that neither in Europe nor in this country are more recent figures available.

As mentioned above, a large part of the improvement is caused by better conditions outside the factories. We included these figures in order to have a complete picture.

It is unfortunate that the excellent British statistics on occupational mortality cover only a few occupations in such a way that they can be used for historical studies. The classification of the groups, especially of those interesting to us, has sometimes changed. The groups in iron foundries, for example, have changed. Painters were sometimes included with decorators but earlier—as in 1900—were classed with plumbers and glaziers. The purpose seems to have been to study special groups. But this change makes impossible in most cases any historical studies of the health hazards of the occupations which would be interesting to us. And even if such studies were possible, it would be very difficult to separate the effects of circumstances outside and inside the factory.

Nevertheless we are able to show clearly and statistically the results achieved by two branches of industrial hygienic activity, namely, the prevention of accidents and occupational diseases. The figures concerning such industrial accidents and occupational diseases as have been reported or compensated according to the laws demonstrate the effect of labor protection in both these fields during the last decades in Europe and, in part, in the United States also. As for accidents, the completeness of the reports varies greatly from country to country. This is due in part to differences in the laws but more largely to the exactitude with which accident reports are made in actual practice. There can be no doubt that the longer the compensation laws have been in effect, the more complete these reports have become. An ever-increasing number of lesser accidents have been reported, whereas the reports of fatal accidents have been much more exact from the beginning.

In Germany, accident-compensation laws were introduced in 1884 and 1887. Statistics show that the following percentage of all accidents resulted in fatalities:

	In Industry	In Agriculture
1888	24.91	43.81
1891	12.85	11.12
1902	7.98	4.85

It is evident that this decline in the percentage of fatal accidents cannot be explained by a quick reduction of the relative frequency of fatal cases but only by an increase in the number of accidents reported, especially less serious ones. Even today it is questionable whether, in every country and every industry, lesser accidents are conscientiously reported. We may assume, however, that almost all fatal accidents are being reported in the most civilized countries. Therefore only the statistics on fatal accidents can be used to prove the effectiveness of accident prevention measures, as was done above in the study of mine accidents.

Statistics on fatal accidents, based on data which the insurance carriers in Germany and Great Britain were required by law to submit to the authorities, were not compiled until several years after the introduction of compensation laws. We have no exact reports covering previous times. Some indication of the earlier rates is to be found in a statement made by Robert Peel. Peel told the House of Commons in 1815 that it was "gratifying to learn that the loss of life had not exceeded one percent per year of those employed in cotton mills" and that this loss "fell short of the average loss sustained in every other class of manufacturing industry." [3]

Unfortunately not all the statistics published in Britain or Germany are available here, but those which we were able to obtain suffice to show the evolution clearly.

Every year until 1938 the English Home Office published statistics of compensation and of proceedings under the Workmen's Compensation Act of 1906 and the Employers' Liability Act of 1880. These reports contain data on seven great groups of industries: mines, quarries, railways, factories, docks, constructional work, and shipping. In Table 7 we show the rates of fatal accidents in all these industries,

TABLE 7

GREAT BRITAIN: FATAL ACCIDENTS PER 1,000 WORKERS

	All Industries	Factories
1911–1914	0.533 [a]	
1919–1924	0.375	0.166 [b]
1925–1929	0.367	0.149
1930–1934	0.327	0.129
1935–1938	0.320	0.123

[a] In 1911 and 1914 great mine disasters occurred. The rate for 1912 and 1913 was 0.497.
[b] For the years 1922–1924.

with factory rates listed separately. Figures cover a period of several years because this procedure gives a clearer picture than do the figures for single years, which sometimes vary widely with the employment of greater numbers of untrained workers and of women and with other factors. The continual decrease in the frequency of fatal accidents is clearly shown.

Statistics available here on Germany cover only the years 1924 to 1939 and seven earlier years not especially selected. Unfortunately, complete, absolute figures are not available, even for the period from 1924 to 1939, so we were not able to establish a table with five-year groups. More-

[3] *Annual Report of the Chief Inspector of Factories for 1943* (London, 1944), p. 15.

over, it should be remembered that since 1926, by an extension of the law, accidents occurring en route to and from work also have been compensable. The actual decrease in accidents is greater than is shown in Table 8 (except in the parentheses). German statistics give, in addition to the proportional figures for 1,000 workers, those for 1,000 *Vollarbeiter,* that is, men who worked 300 days, but the differences as a result of this method are small. Figures given in Table 8 include those of all the *gewerblichen Berufsgenossenschaften* (industrial insurance associations), including mines and transportation, and also those of three special groups of industries which have different rates of fatal accidents. In Germany and England in the various industries, the tables show a remarkable diminution in fatal accidents. A very small part of this diminution in Great Britain, as in Germany, may have been caused by the progress of surgery, but it is due far more to the progress of accident prevention.

For the United States as a whole, reliable figures are available only on accidents in coal mines and other mines. These figures, some of which are shown in Table 6, were compiled by the United States Bureau of Mines. But there are no such complete statistics available on accidents in other industries. This is due to the division into many states, each with its own legislation and institutions. In the single states, accident insurance usually is provided by many insurance companies, which do not necessarily report figures relating to accidents and even less do they report the number of insured workers.

The Bureau of Labor Statistics, especially M. D. Kossoris, later head of the Division of Accident Statistics, has tried for some years to collect such data concerning industry as a whole and separate industries. These statistics are based on voluntary reports of the establishments, and began in the middle twenties with reports from 6,000 or 7,000 estab-

Table 8
Germany: Fatal Accidents

	PER 1,000 WORKERS				PER 300,000 WORKING DAYS			
Year	All Industries, Mines, Transportation	Iron and Metals: Extraction, Manufacture, Apparatuses	Chemical Industry	Textile Industry	All Industries	Iron and Metals	Chemical Industry	Textile Industry
1897	0.824	..	..	..	..	..	..	..
1900	0.834	..	..	..	..	..	..	..
1902	..	..	..	0.08	0.76	..	0.62	0.10
1903	0.62	..	..	..	0.72	..	..	0.12
1905	0.710	..	..	..	..	..	..	..
1911	..	..	..	..	0.68	..	0.64	0.11
1913	0.62	0.57	0.50	0.08	0.69	..	0.51	0.08
1924	0.45	0.34	0.52	0.08	0.51	0.36	0.52	0.09
1925	0.49	0.37	0.49	0.08	0.54	0.39	0.49	0.09
1926	0.49 [a]	0.40	0.51	0.09	0.56	0.43	0.51	0.11
1927	0.47	0.36	0.45	0.08	0.52	0.38	0.46	0.09
1928	0.48	0.42	0.51	0.09	0.54 (0.45+) [a]	0.44	0.51	0.10
1929	0.40	0.34	0.36	0.07	0.46 (0.43)	0.36	0.36	0.07
1930	0.40	0.30	0.28	0.06	0.45 (0.426)	0.33	0.28	0.07
1931	0.33	0.26	0.25	0.06	0.38 (0.37)	0.28	0.25	0.06
1932	0.28	0.22	0.25	0.06	0.32 (0.292)	0.25	0.25	0.06
1933	0.27	0.22	0.25	0.05	0.31 (0.285)	0.24	0.25	0.06
1934	0.29	0.23	0.23	0.05	0.33 (0.293)	0.25	0.23	0.05
1935	0.31	0.26	0.33	0.05	0.35 (0.312)	0.28	0.33	0.06
1936	0.31	0.27	..	0.05	0.35	0.22	..	0.05
1938	..	..	..	..	..	0.27	0.26	0.05
1939	..	..	..	..	0.37	0.28	0.29	0.05

[a] Since 1926, accidents occurring on the way to and from work have been compensated and therefore are included here. The number exclusive of these accidents is shown in parentheses.

THE EFFECTS OF INDUSTRIAL HYGIENE 273

lishments in about 30 manufacturing industries. Since 1936 the comparison between two consecutive years has been based on reports of about 20,000 identical establishments reporting in both years. The figures of the United States census cover all plants whose products were valued for the year at $5,000 or more. If we assume that no smaller establishments reported accident statistics and therefore compare both figures—accident reports and census—we find that reports were submitted by 11.5 percent of all the establishments, employing 44 percent of all the persons engaged in manufacture. But for chemicals, iron and steel, and textiles, reports were submitted by 15.9, 25.5, and 38.5 percent of the establishments respectively, with 77, 76.5, and 61 percent of the employees in these branches. So we may conclude that preponderantly large factories reported. Conclusions drawn from figures received in this way can be used only with certain reservations.

M. D. Kossoris and S. Kjaer, who for years compiled these statistics, which were published in the *Monthly Labor Review* and in special bulletins of the Bureau of Labor Statistics, tried by adjustment to correct the statistics of earlier years. Nevertheless these figures, based on such a small number of reports in the first years, do not seem reliable even when corrected in accordance with later experiences. The authors calculate that if we put the frequency rate of death and permanent total disability in 1926 at 100, the rate in 1937 would be 85.7, in 1938, 71.4, and would reach 62.8 in 1944. I am inclined to believe that the latest proportion only, 71.4 to 62.8, has a certain significance. But I have no intention of following the estimates of accident statistics made by these authors for the whole of American industry. I believe also that the figures on "total manufacture," that is, all the factories reporting, have a lower value than those for the single industries because the different industries that make up this "total" may not be identical from year to year.

If we calculate from the material amassed by these authors the frequency rate of fatal accidents per one million working hours for several recent years, we arrive at the figures shown in Table 9. These rates indicate first a

TABLE 9

UNITED STATES: RATE OF FATAL ACCIDENTS PER 1,000,000 WORKING HOURS

	Total Manufacture	Chemical Products	Iron and Steel Products	Textile Products
1937	0.103	0.160	0.133	0.029
1938	0.079	0.122	0.122	0.029
1939	0.076	0.126	0.124	0.029
1940	0.095	0.137	0.130	0.093
1941[a]	0.095	0.137	0.113	0.024
1944	0.093	0.168	0.144	0.045

[a] Approximately calculated.

decline, then the fluctuations such as are always found in accident statistics, but on the whole no tendency toward a decrease. It must be remembered that war years are often exceptional, and that in a period of so few years, some of them during wartime, a marked decline cannot be expected.

Exact data may be available in those states in which insurance in a state fund is compulsory or where other insurance is allowed only in exceptional cases. But such states are few, and among them are states too small to provide a useful picture. West Virginia is the only one which compiles figures valuable for the purposes of this study. The accident fatality rates per 1,000 workers in West Virginia industries, exclusive of coal mining, are as follows:

Year	Rate
1924–25	1.410
1934–35	0.676
1939–40	0.498

THE EFFECTS OF INDUSTRIAL HYGIENE   275

Year	Rate
1941–42	0.495
1944–45	0.463
1945–46	0.475

These figures show a clear and important decrease in fatal accidents in the industry of this state. In view of these statistics, and bearing in mind that the Bureau of Labor Statistics believes that its calculations indicate a decrease in fatal industrial accidents in the country as a whole, we may agree that in the United States accident rates seem to be declining somewhat.

An exact comparison between the figures for England, Germany, and the United States is not possible, because we do not have the records of separate industries in England and also because the German statistics class as fatal accidents all those accident cases which ended in death within the subsequent year. Comparing the American figures with the European, we find the figures of West Virginia much higher than those of Great Britain.

In comparing the figures of the Bureau of Labor Statistics with the German figures, we must translate the German figures for 300 workdays of 1,000 workers into working hours. If it is justifiable to calculate the working hours in Germany in the years before the war on the basis of an eight-hour day, 1,000 *Vollarbeiter* would work 2,400,000 hours a year. Thus for the year 1939, the latest for which figures are available from Germany, we arrive at the following rates:

FATAL ACCIDENTS PER ONE MILLION WORK HOURS

Industry	United States	Germany
Chemical products	0.126	0.121
Iron, Steel, and products	0.124	0.117
Textiles	0.029	0.021

In the year 1939 the German figures were higher than those of the preceding years, but it was a favorable year for the

United States. Nevertheless, the rates were higher in this country than in Germany.

Perhaps it would be more correct to calculate for this year, in which Germany prepared for war, more working hours per day. Estimating the workday as ten hours, the German figures per million work hours would be even smaller: 0.097, 0.093, and 0.017. The figures from West Virginia, although they do not include mines, are much higher than the German figures (Table 8) which include mines.

To obtain an accurate picture of occupational diseases we cannot use the statistics on fatal cases, because these figures are too small and also because the diagnoses in these cases are neither reliable nor complete. However, the decline in the number of diseases, particularly in the poisonings, is so great that it is convincing, even if the figures are not entirely reliable. Table 10 is based on the Annual Reports of the Chief Inspector of Factories of Great Britain. It concerns cases "reported" by the practitioner, according to law. There are incorrect diagnoses here also, but the decline is greater than any probable change in diagnosis could account for. The number of cases compensated for lead poisoning (Table 11), in which the diagnoses have been established as exact, also shows a great decline.

TABLE 10

GREAT BRITAIN: REPORTED CASES OF SEVERAL OCCUPATIONAL DISEASES, BY YEARS [a]

	1900	1910	1920	1930	1935	1940	1944
Lead Poisoning	1,058	505	289	265	168	108	41
White and red lead works	377	44	28	3	7	17	2
Potteries	210	78	25	23	18	6	..
Mercurial poisoning	9	10	5	3	1	..	1
Arsenical poisoning	22	7	3	1	1	2	3
Anthrax	37	51	48	43	20	37	8
Anilin poisoning	..	..	..	24	9	64	55

[a] Annual Reports of the Chief Inspector of Factories.

TABLE 11

GREAT BRITAIN: COMPENSATED CASES OF LEAD POISONING, BY YEARS OF FIRST PAYMENT [a]

Year	Cases	Year	Cases
1908	421	1925	234
1910	367	1930	176
1920 [b]	169	1935	143

[a] The figures for 1908 to 1930 are taken from T. M. Legge, *Industrial Maladies;* those for 1935 are from *Workmen's Compensation Statistics.*
[b] Influenced by World War I.

It may be pointed out that the war years influenced the frequency of incidence of some occupational diseases. Anthrax decreased because of the greatly diminished imports of wool from abroad. On the other hand, war production of ammunition considerably increased the manufacture and use of derivatives of benzol and its homologues, with a consequent increase in the cases of poisoning caused by them. In 1930, 24 cases were reported under the heading of "anilin" poisoning; these increased to 249 in 1941 and to 204 in 1942, but were quickly reduced to 79 in 1943.

Table 12 shows the number of compensated cases in Germany, the diagnoses of which were approved by experienced medical men. The increase in the first years is explained by the fact that some time elapsed before the knowledge of possible compensation was widespread.

TABLE 12

GERMANY: COMPENSATED CASES OF OCCUPATIONAL POISONING, BY YEARS OF FIRST PAYMENT

	1926	1930	1931	1932	1934	1935	1936	1937	1938	1939
Lead	241	470	371	198	116	..	..	..	83	114
Mercury	13	13	6	11	9	..	..	..	13	13
Carbon Disulphide	5	2	10	1	9	1	..	..	9	9

Table 13 is reproduced from the reports of the Association of Viennese Sickness Insurance Funds (compulsory insurance). The figures shown here cover only a single district—Vienna. In 1905 the Association appointed a physician to treat occupational diseases and try to reduce the number of cases. With the cooperation of the unions, the factory inspectors, and the government, the results shown in the table were realized. The following facts help to explain the findings. In 1906 a strike of filecutters gave the workers an opportunity to request and push through the replacement of lead in the supports by a lead alloy. A rule issued by the Minister of Commerce in the same year prohibited the dyeing of silk yarns with lead compounds, a practice which had endangered dyers and haberdashers.

TABLE 13

VIENNA: WORKERS ILLNESSES WITH DISABILITY, CAUSED BY LEAD POISONING

Year	File Cutters	Haberdashers	Silk Dyers	Painters	Factory Workers (Women)
1902	13	4	5	125	85
1903	14	4	8	163	80
1904	21	5	20	197	120
1905	17	8	20	198	85
1906	11 [a]	16 [b]	7 [b]	252	124
1907	9	9	5	208	112
1908	7	1	1	167 [c]	87
1909	8	1	2	143	52
1910	4	2	0	138	34
1911	4	0	?	110	12
1912	9	1	0	116	3

[a] Agreement (1906) to replace lead supports with others made from lead alloy.
[b] Rule of July 17, 1906, forbidding the use of lead compounds in dyeing silk.
[c] Rule of April 15, 1908, prohibiting the use of lead compounds in painting the interior of buildings.

A regulation of 1908 forbade the use of lead paints in the interior of buildings, thus protecting most of the painters. The use of lead paint for the outside of buildings is not usual in Vienna except for iron construction. Lead poisoning among women industrial workers occurred, with few exceptions, only in a plant manufacturing metal bottle caps. The energetic pressure brought to bear by the authorities, to whom the information was delivered by the sick fund and by physicians, eliminated this source of lead poisoning.

The whole procedure in the trades mentioned here is a good example of cooperation in which the government and the unions and sick funds played the most important roles.

Although it is not possible to show the effects of factory hygiene in general, the beneficial results in the prevention of accidents and occupational diseases as seen in the examples of Great Britain, Germany, and Vienna are, I think, sufficiently strong support for the view that the other branches of factory hygiene have not worked in vain either; they, too, have played a part in the improvement of health and the lowering of mortality as shown in the statistics.

All the European tables, reaching far back, point to one fact: at the time when the government began to exercise vigorous control over the circumstances leading to accidents or occupational diseases, the rates of incidence for both were high; they were quickly brought down. This indicates that neither private initiative nor the good will of individual managements have been able to improve conditions so thoroughly as has governmental activity. Detailed regulations and the enforcement thereof were necessary to bring about great improvement. The same fact seems to be indicated by the statistics of accidents and especially of mine accidents in the United States in comparison with those of other countries.

13

Summary

THE INFORMATION on industrial hygiene in antiquity is very scanty. It is true that many physicians since Hippocrates have told us something about occupational diseases, especially lead poisoning, and also about their treatment. But very little is said of their control, and some of what is said—for instance the advice of Pliny on protection in breathing—offers an inefficient method.

During the Middle Ages nothing new was added, which is easily explained by the fact that all industrial activity was performed in small workshops. The mines also were small. There were only a few exceptions of larger workshops and mines. The low level of the natural sciences, of medicine, and technology made an efficient control impossible.

The first efforts to diminish industrial hazards by technical means we find in the mines, where without such installations the work had become impossible (Agricola).

Although the work of Ramazzini (1700) does not contain much about industrial hygiene, the works of some of his translators and revisers (Ackermann, Patissier) show great progress, seemingly due to the influence of the encyclopedists and the French Revolution. The beginning of the development of technology also was followed by an increasing number of prophylactic measures.

The economic evolution, which in England in the 18th century initiated the rise of capitalism and the use of machinery, created horrible conditions inside factories as well as outside in the workers' districts. England, which had taken the lead in the development of industry, also took the lead in the corresponding labor protection and industrial hygiene.

SUMMARY

In all countries the first labor laws were laws for the protection of children, sometimes embodying regulations related to measures of hygiene. The enactment of such laws was preceded by severe struggles between—on the one hand—manufacturers, the economic school of laissez faire, laissez aller, and the dogma of free enterprise without government interference and—on the other hand—the humanitarians (often from the class of manufacturers themselves), intelligent governmental officials, and the growing influence of the working class. These struggles finally resulted in the enforcement of protective laws and the appointment of special officials invested with appropriate powers and finally the investment of governmental employees with the right to issue—under parliamentary control—regulations for hazardous industries.

The injuries to human life and health in the first ruthless years of industrialization were so evident that it did not at first require scientific research to show the necessity of improvements and to recognize the remedies. But very soon it proved necessary to investigate working conditions and their effect on human health. This was done by private individuals, scientists and physicians, and shortly thereafter in England by governmental councils and officials as well—particularly factory inspectors—appointed for the enforcement of the early labor laws. A little later such investigations also were undertaken in the other European countries. Thus there developed in the third and especially in the fourth quarter of the 19th century a science of industrial hygiene. It developed further, using the methods available at that time in the medical, technical, and statistical sciences.

The growing realization that men are more important than economy, that the latter must serve the well-being of the people, that the prosperity of both employer and employed are interrelated—these and the increasing influence of the working class promoted the practical use of the scientific and technical knowledge acquired and spurred the hygienic sciences to proceed still further. When the most easily recognized evils had been at least partially removed, the recognition of others—and hence their control—could be achieved only by exact studies. Thus the science of industrial hygiene widened and deepened.

Roentgenology made us acquainted with dust diseases, first studied by this means in South Africa; the methods of dust counting, developed there and improved in the United States, showed us the way to their control. The clinical studies of occupational diseases in England and Germany gave us the basis for hygienic measurements, medical (especially periodic examination) and technical. Industrial science developed ventilation and exhaust devices. The elaboration of the finest chemical methods, performed primarily in the United States, is of further help and brings new hope for the future.

The task remaining is first of all to put into more widespread practice all the measures discovered in these ways, to restrict dangers and control noxious substances, and to improve industrial health generally. The legislation in England and Germany and the efficiency of factory inspection in these two countries point the way. Numerous laws—exact and detailed rules and regulations—are necessary, as well as their enforcement by highly qualified factory inspectors.

As far as it is possible to evaluate the situation from available statistics, we find a high incidence of accidents and occupational diseases at the beginning of state intervention, followed by a distinct decrease in consequence of this intervention. The better the governmental regulations and the more efficiently they are enforced, the greater has been the actual progress in eliminating accidents and diseases. These laws and regulations together with the officials for their enforcement, the factory inspectors, are the backbone of practical industrial hygiene in Europe and in several states of the United States.

In this country—where industry is younger but has outranked the European in technique,—the introduction of laws, rules, and regulations and provisions for their enforcement began much later and are much less developed in many of the states than in Europe. Although an immense amount of research work is done by government institutions and continual efforts are made for "education," the results, as far as can be recognized by the statistics, are behind those achieved in Europe.

All of which shows that laws and regulations, thoroughly worked out and enforced by well-trained inspectors, are indispensable to the practice of industrial hygiene. With these as a basis, it is important to have the cooperation of associations of experts, of trade unions, scientists and physicians.

Bibliography

OFFICIAL SOURCES

Austria

Ministry of the Interior. Krankheits- und Sterblichkeitsverhältnisse bei den . . . Krankenkassen, 1896–1910. Vienna, 1913.

K. K. Arbeitsstatistisches Amt im Handelsministerium. Bleivergiftungen in Hüttenmännischen und Gewerblichen Betrieben. Parts I–IX. Vienna, 1905–1915.

France

Ministère du Commerce de l'Industrie, Office du Travail. Poisons industriels. 1901.

Germany

Preussisches Ministerium für Handel und Gewerbe. Grubensicherheitsamt. "Die Bekämpfung des Bohrstaubes im Bergwerksbetrieb: Bericht über das Ergebnis des 'Preisausschreibens für Bohrstaubschutz.'" Zeitschrift für das Berg-Hütten- und Salinenwesen im Preussischen Staate, Vol. LXXIX, 1931.

Kaiserliches Statistisches Amt. Krankheits- und Sterblichkeitsverhältnisse in der Ortskrankenkasse für Leipzig und Umgebung. 1910.

Reichsamt des Innern. Die Arbeiterschutzvorschriften im Deutschen Reich. 1915.

Great Britain

Chief Inspector of Factories. Annual Report. Home Office, up to 1938; Factory Department of the Ministry of Labour and National Services, from 1939.

Chief Inspector of Factories and Workshops. Annual Report for

the Year 1932, Including a Review of the Years 1832–1932. 1933.
Committee on the Health of Cornish Metalliferous Miners. Report. Home Office, 1904.
Departmental Committee Appointed to Inquire into the Dangers . . . of Lead . . . and Dust and Other Causes in the Manufacture of Earthenware and China. Report, Vols. I–III. Home Office, 1910.
Departmental Committee Appointed to Inquire into the Law Relating to Compensation for Injuries to Workmen. Vol. I: Report; Vol. II: Minutes of Evidence. Home Office, 1904.
Departmental Committee on Humidity and Ventilation in Cotton Weaving Sheds. Report, Vols. I–II. Home Office, 1909, 1911.
Department of Scientific and Industrial Research. Methods for the Detection of Toxic Gases in Industry: Hydrogen Bisulfide (1937); Hydrogen Cyanid, Sulfur Dioxid (1938); Benzol, Nitrous Fumes, Chlorine (1939); Organic Halogen Compounds (1940).
Englischen Frabriksgesetze, Die. Tr. by Benno Karpeles. Berlin, 1900.
Lords Commissioners of the Admiralty, Committee Appointed to Consider and Report upon the Conditions of Deep-Water Diving. Report. 1907.
Medical Research Council. Chronic Pulmonary Disease in South Wales Coal Miners. Special Reports 243, 244, 250. Privy Council, 1942, 1943, 1945.
—— Industrial Fatigue Research Board, Reports. Nos. 1–2 (1919); Nos. 3–13 (1921); Nos. 14–54 (1921–1929); No. 55 (1930).
—— T.N.T. Poisoning and the Fate of T.N.T. in the Animal Body. Privy Council, 1921.
—— Report of the Miners' Nystagmus Committee. Privy Council, 1922, 1923, 1932.
Mines Department. Annual Reports of the Secretary of Mines and of the Chief Inspector of Mines.
Ministry of Munitions, Health of Munition Workers Commit-

tee. Industrial Efficiency and Fatigue: Interim Report, 1917; Final Report, 1918.
National Health Insurance. The Causation and Prevention of Tri-nitro-toluene (T.N.T.) Poisoning. 1917.
Privy Council. Third and Fourth Reports of the Medical Officer. 1860, 1862.
The Registrar-General's Decennial Supplement: Occupational Mortality. Home Office, decennially from 1861.
Royal Commission Appointed to Inquire into the Conditions of the Mines in Great Britain. Report. 1864.
Royal Commission on Metalliferous Mines and Quarries. Second Report. 1914.
Statistics of Compensation and of Proceedings under the Workmen's Compensation Acts. Home Office, 1909-1938.

NETHERLANDS

Bijdragen tot de Statistiek van Nederland, No. 247: Statistiek van de Sterfte onder de mannen; met underscheiding naar beroep. s'Gravenhage, 1917.

UNION OF SOUTH AFRICA

Miners' Phthisis Prevention Committee. General Report, Pretoria, 1916; Final Report. Pretoria, 1919.
Miners' Phthisis Board. Report, 1940-1941.
Report upon the Work of the Miners' Phthisis Medical Bureau. Pretoria and Cape Town, 1923-.

UNITED STATES

Bureau of Mines. Approval Schedule 21: Procedure for Testing Filter-Type Dust, Fumes, and Mist Respirator for Permissibility. 1934.
—— Approval Schedule 14E: Procedure for Testing Gasmasks for Permissibility. 1941.
—— Coal Mine Accidents for 1941. Bulletin 456.

Bureau of Mines. Federal Mine Safety Code for Bituminous Coal and Lignite Mines of the United States. 1946.
—— Minerals Yearbook.
Department of Labor. Annual Digest of State and Federal Labor Legislation. Division of Labor Standards, yearly.
—— Handbook of Federal Labor Legislation. Division of Labor Statistics, Bulletin 39: Part I (1940), Part II (1941).
—— Qualifications for General Labor Law Inspectors, Prepared by an Advisory Committee. Division of Labor Standards, Bulletin 38. 1940.
—— State Workmen's Compensation Laws as of June, 1946. Division of Labor Standards, Bulletin 78.
—— Statistics of Industrial Accidents in the United States to the End of 1927. Division of Labor Statistics, Bulletin 490.
—— Work Injuries in the United States during 1944. By M. D. Kossoris. Bureau of Labor Statistics, Bulletin 849.

GENERAL SOURCES

Ackermann, Johann Christian Gottlieb. B. Ramazzini's Anhandlung von den Krankheiten der Künstler und Handwerker. New ed., rev. and enl. Stendal, 1780.
Actuarial Society of America and the Association of Life Insurance Medical Directors. Joint Occupation Study, 1928. New York, 1929.
—— Occupational Study, 1937. New York, 1938.
Adelmann, Georg. Ueber die Krankheiten der Künstler und Handwerker. Würzburg, 1803.
Agricola, Georg. De re metallica, libri xii. Basel, 1556. English ed., tr. by H. C. and L. H. Hoover, London, 1912. New German ed., *Zwöl Bucher vom Berg-und Huttenwesen*, Berlin, 1928.
Alberti, Michael. De preferendis metallicorum morbis. Halle, 1721.
Albrecht, H., ed. Handbuch der praktischen Gewerbehygiene mit besonderer Berücksichtigung der Unfallverhütung. Berlin, 1896.

BIBLIOGRAPHY

Alwens, W., E. E. Bauke, and W. Jonas. "Auffallende Häufung von Bronchialkrebs bei Arbeitern der chemischen Industrie." *Archiv fur Gewerbepathologie*, VII (1937), 69–84.

American Standards Association. American Safety Standard Codes. New York, 1924–.

Ammann, Jost. Eigentliche Beschreibung aller Stände mit Reimen von Hans Sachs. Frankfort on Main, 1568. New ed., Leipzig, c. 1910.

Anderson, Adelaide M. "Historical Sketch of the Development of Legislation for Injuries and Dangerous Industries in England." Chapter II in *Dangerous Trades*, ed. T. Oliver. London, 1902.

Andrae, Albert. "Die Sterblichkeit in den Berufen, die sich mit der Herstellung und dem Verkauf geistiger Gertränke befassen." *Zeitschrift fur die gesammte Versicherungs-Wissenschaft*, 1905.

—— "Die Sterblichkeit in den land- und forstwirtschaftlichen Berufen." *Zeitschrift für die gesammte Versicherungs-Wissenschaft*, 1906.

Andrews, John B. "Phosphorus Poisoning in the Match Industry in the United States." Bureau of Labor, Bulletin 86. Washington, 1910.

—— "Deaths from Industrial Lead Poisoning (Actually Reported) in New York State in 1909 and 1910." In Bureau of Labor, Bulletin 95 (Washington, 1911), 260–282.

—— Administrative Labor Legislation. New York, 1936.

—— Labor Laws in Action. New York, 1938.

Arlidge, J. T. The Hygiene, Diseases and Mortality of Occupations. London, 1892.

Arnold, S. Untersuchungen über Staubinhalation und Staubmetastase. Leipzig, 1885.

Assiette au Beurre. Paris, April 5, 1905. (Concerning lead paints.)

Aub, J. C., L. T. Fairhall, A. S. Minot, and P. Reznikoff. Lead Poisoning. Baltimore, 1926.

Automobile Workers. See United Automobile Workers.

Baader, E. W., and E. Holstein. Das Quecksilber und die gewerbliche Quecksilbervergiftung. Berlin, 1933.

Badham, C. Reports of the Medical Officer of Industrial Hygiene; Extracts from the Reports of the Director-General of Public Health, New South Wales, for the Years 1923–1944. Sydney, Australia.

Badham C., and H. B. Taylor. The Lungs of Coal, Metalliferous and Sandstone Miners in New South Wales. Sydney, Australia, 1938.

Baer. "Ueber den Nystagmus der Bergleute." *Deutsche Medizinische Wochenschrift*, II (1876), 147–150.

Ballard, J. W., H. H. Schrenk, and H. I. Oshry. Quantitative Analysis by X-ray Diffraction, I: Determination of Quartz. Bureau of Mines, Report 3520. Washington, 1940.

Bartels, "Tagung der Betriebs- und Fabrikärzte . . . 1936." *Zentralblatt für Gewerbehygiene*, XXIII (1936), 107–108.

Bassoe, John J. "The Late Manifestation of Compressed Air Disease." *American Journal of Medical Sciences*, CXLV (1913), 129–152.

Bauer, Stephan. Gesundheitsgefährliche Industrien. Jena, 1903.

—— Die gewerbliche Nachtarbeit der Frauen. Jena, 1903.

Beck, K., and P. Stegmüller. "Ueber die Löslichkeit von Bleisulfat und Bleichromat." *Arbeiten aus dem kaiserlichen Gesundheitsamt*, Vol. XXXIV (1910).

Becourt, and A. Chevallier. "Memoire sur les accidents, atteignent les ouvriers, qui travaillent la bichromate de potassium." *Annales d'hygiène publique et de médicine légale*, 2d ser., XX (1863), 83–95.

Behnke, A. R., et al. "The Rate of Elimination of Dissolved Nitrogen in Relation to the Fat and Water Content of the Body." *American Journal of Physiology*, Vol. CXIV (1935–1936).

—— "Employment of Helium in Diving to New Depths of 440 Feet." *United States Naval Medical Bulletin*, XL (1942), 65.

Bering, E. and E. Zitzke. Berufliche Hautkrankheiten. Leipzig, 1935.

Bibra, von, and Geist. Die Krankheiten der Arbeiter in den Phosphorzündholzfabriken. Erlangen, 1847.

Blänsdorf, E. "Bleiliteratur." *Schriften aus dem Gesammtgebeit der Gewerbehygiene*, Vol. II, No. 7 (Berlin, 1922).

Blaschko, A. "Gewerbliche Hautkrankheiten." *Handbuch der Arbeiterkrankheiten*, ed. T. Weyl. Jena, 1908.
Bleicher, H. *See* Frankfurter Krankheitstafeln.
Bloch, B. "Aetiologie und Pathogenese des Eczems." *Archiv für Dermatologie*, CLIV (1924), 34–82.
Bloomfield, J. J., J. M. Dallavalle, R. R. Jones, and Others. Anthracosilicosis. Public Health Bulletin 221. Washington, 1936.
Bloomfield, J. J., and L. Greenburg. "Sand and Metallic Abrasive Blasting as an Industrial Health Hazard." *Journal of Industrial Hygiene*, XV (1933), 184–204.
Bloomfield, J. J., V. M. Trasko, and Others. A Preliminary Survey of the Industrial Hygiene Problem in the United States. Public Health Bulletin 259, Washington, 1940.
Blyth, A. W. Poisons, Their Effects and Detection. 1st ed., New York, 1885; 5th ed., with M. W. Blyth, London, 1920.
Böhme, A. "Zur Kenntnis des Röntgenbildes der Lungenanthrakose." *Fortschritte auf dem Gebiete der Röntgestrahlen*, XXIX (1922), 301–311.
—— "Die Pneumokoniose der Bergarbeiter im Ruhrbezirk." *Fortschritte auf dem Gebiete der Röntgenstrahlen*, XXXIII (1925), 39–50.
Boisier de Sauvages. Nosologie méthodique. Lyon, 1772.
Bornstein and Plate. "Ueber chronische Gelenksveränderungen enstanden durch Presslufterkrankung." *Fortschritte auf dem Gebiete der Röntgenstrahlen*, XVIII (1911–1912), 197–206.
Boulin, M. Les Fonderies de plomb. Paris, 1907.
Bowditch, M., C. K. Drinker, H. H. Haggard, and Alice Hamilton. "Code for Safe Concentrations of Certain Common Toxic Substances Used in Industry." *Journal of Industrial Hygiene and Toxicology*, XXII (1940), 251.
Bowditch, M., and H. B. Elkins. "Chronic Exposure to Benzene, I: The Industrial Aspects." *Journal of Industrial Hygiene and Toxicology*, XXI (1939), 321–333.
Boyd, R. N. Coal Mines Inspection; Its History and Results. London, 1879.
Breton, J. L. Rapport au nom de la Commission de l'Hygiène publique chargés d'examiner le projet de loi . . . sur l'em-

ploi des composées du plomb dans les travaux de la peinture de batiments. Chamber of Deputies, No. 799. Paris, 1907.

Brezina, E. "Ueber die Wirkung der gebräuchlichen Respiratoren." *Archiv für Hygiene*, LXXIV (1911), 143–163.

—— Internationale Uebersicht über Gewerbekrankheiten, nach den Berichten der Gewerbeinspectionen der Kulturländer. For the years 1909–1912, published as a supplement to the *Oesterreichischen Sanitätswesen* (Vienna, 1913–1916); for 1913–1929, in *Schriften aus dem Gesammtgebiet der Gewerbehygiene* (Berlin, 1921–1931).

Brockmann, C. H. Die metallurgischen Krankheiten des Oberharzes. Osterode, 1851.

Bronce, F. A. Bibliography and Survey of Lead Poisoning. New York, 1943.

Brouardel, Paul. Les Empoisonnements criminels et accidentels. Paris, 1902.

—— Les Intoxications. Paris, 1904.

Brown, J. Health; Five Lay Sermons to Working People. New York, 1865.

Browning, Ethel. Toxicity of Industrial Organic Solvents. London, 1937.

Brundage, Albert H. Manual of Toxicology. 1st ed., Brooklyn, N.Y., 1911; 15th ed., New York, 1926.

Brundage, Dean K. "Sickness among Persons in Different Occupations of a Public Utility." *Public Health Reports*, Vol. XLIII (1928).

—— "A Survey of the Work of Employees of Mutual Benefit Associations." *Public Health Reports*, Vol. XLVI (1931).

Bruns, H. "Durch Eingeweidewürmer bedingte Berufskrankheiten." *Handbuch der sozialen Hygiene*, Vol. II (Berlin, 1926).

Bubbe, Joannes. De spadone hippocratico lapicidorum Seebergensium haemoptysin et phthisin pulmonum (vulgo die Seeberger Steinbrecherkrankheit). Halle, 1742.

Burke, W. J., S. Moskowitz, J. Siegel, B. H. Dolin, and C. B. Ford. Industrial Air Analysis, a Description of Some of the Chemical Methods Employed in the Laboratory. Division of Industrial Hygiene, New York State Dept. of Labor, 1948.

Büttner-Wobst, W., and O. Trillitzsch. "Die Bergflachslunge (Asbestosis) und was der deutsche Arzt von ihr wissen muss." *Tuberkulose*, XI (1931), 11–14.

Calder, J. Prevention of Factory Accidents. London, 1899.

Carnot, P., E. Lancereaux, M. Letulle, and R. Würtz. Intoxications. Paris, 1907.

Casper, J. L. Die wahrscheinliche Lebensdauer. Berlin, 1835.

Castellino, N. Il Lavoro nella chimica industriale. Milan, 1940.

Christ, A. "Ueber Caissonkrankheit mit besonderer Berücksichtigung einer typischen Erkrankung des Hüftgelenkes." *Deutsche Zeitschrift für Chirurgie*, CCXLIII (1934), 132.

Cless. "Beiträge zu einer Krankheitsstatistik der Gewerbe." *Haesers Archiv für die gesammte Medizin*, III (1842), 258–274.

Collis, E. L. Special Report on Dangerous or Injurious Processes in the Smelting of Material Containing Lead and in the Manufacture of Red and Orange Litharge and Flaked Litharge. London, 1910.

—— "The Effects of Dust in Producing Diseases of the Lungs." In *XVIIth International Congress of Medicine, Section on Hygiene and Preventive Medicine*. London, 1913.

—— Industrial Pneumonoconioses (Milroy Lectures, 1915). London, 1919.

Collis, E. L., and Major Greenwood. The Health of the Industrial Worker. London, 1921.

Cook, Warren A. "Maximum Allowable Concentrations of Industrial Atmospheric Contaminants." *Industrial Medicine*, XIV (1945), 936–946.

Cooke, W. E. "Fibrosis of the Lungs Due to the Inhalation of Asbestos Dust." *British Medical Journal* (1924), p. 147.

Curschmann, F. "Eignung der Frau zur Arbeit in der Chemischen Industrie." *Zentralblatt für Gewerbehygiene*, VI (1918), 48–53, 61–65.

Dammer, O., ed. Handbuch der Arbeiterwohlfahrt. Stuttgart, 1902.

Davis, G. G., E. M. Salmonsen, and J. L. Earlywine. The Pneumoconiosis (Silicosis); Literature and Laws. Chicago, 1935–1947.

Decondé. "Note sur le nystagmus." *Archives belgique de médicine,* XXVII (1861), 337.

Dejerine-Klumpke, Madame. Des polyneurites en general et des paralysies et atrophies saturnines en particulier. Paris, 1889.

Delmege, J. A. Towards National Health. London, 1931.

Delpech, A. Memoire sur les accidents que développe chez les ouvriers en caoutchouc, l'inhalation du sulfure de carbone en vapeur. Paris, 1856.

Diemerbroeck. Anatome corporis humani. Isbrand, 1683.

Dreessen, W. C. "Effects of Certain Silicate Dusts on the Lungs." *Journal of Industrial Hygiene,* XV (1933), 66–78.

Dreessen, W. C., and J. M. Dallavalle. "The Effects of Exposure to Dust in Two Georgia Talc Mills and Mines." *Public Health Reports,* Vol. L (1925).

Dreessen, W. C., J. M. Dallavalle, T. L. Edwards, R. R. Sayers, and others. Pneumoconiosis among Mica and Pegmatite Workers. Public Health Bulletin 250. Washington, 1940.

Drinker, C. K. Carbon Monoxide Asphyxias. New York, 1938.

Drinker, P., and T. Hatch. Industrial Dust: Hygienic Significance, Measurement and Control. New York, 1936.

Dublin, L. I. Causes of Death by Occupation. U.S. Bureau of Labor Statistics, Bulletin 207. Washington, 1917.

Dublin, L. I., and R. J. Vane. Causes of Death by Occupation. U.S. Bureau of Labor Statistics, Bulletin 507. Washington, 1930.

Ebert, Max. Reallexicon der Vorgeschichte. Berlin, 1924.

Elkins, H. B. "Maximal Allowable Concentrations, I: Carbon Tetrachloride." *Journal of Industrial Hygiene and Toxicology,* XXIV (1942), 233–235.

Elkins, H. B., and L. Levine. "Decomposition of Halogenated Hydrocarbon Vapors by Smoking." *Journal of Industrial Hygiene and Toxicology,* XXI (1939), 221–225.

Ellenbog, Ulrich. Von den giftigen, besen Tempfen und Reuchen der Metal. 1524; new ed. by F. Koelsch and F. Zoepfl, Munich, 1927.

L'Empoisonnement par le plomb. Conference, under the presi-

dency of Paul Brouardel, Paris, January 13, 1901. Report by J. V. Laborde.
Engel, H. "Ueber die Gesundheitsgefährdung bei der Verarbeitung von metallischen Blei mit besonderer Berücksichtigung der Bleilöterei," *Schriften aus dem Gessammtgebiet der Gewerbehygiene,* new ser., No. 13 (Berlin, 1925).
Engel, H., and V. Froboese. "Untersuchungen zur Klärung der Bleiverflüchtigung beim homogenen Verbleien und Bleilöten unter Verwendung verschiedener Gebläseflammen." *Archiv für Hygiene,* XCVI (1926), 69–101.
Erben, F. Vergiftungen: Vol. VII of *Handbuch der Aerztlichen Sachverständigen-Tätigkeit.* Vienna, 1909, 1910.
Erdmann. "Von den Krankheiten der Steinkohlenarbeiter in den Gebirgen des Plauenschen Grundes." *Journal der praktischen Heilkunde,* ed. by Hufeland and Ossan, new ser., LXVI, No. 6 (1831), 3–21.
Escherich. Hygienisch-statistische Untersuchungen über Lebensdauer in verschiedenen Ständen (Geistliche, Lehrer, Beamte) in Bayern. Würzburg, 1854.
Eulenberg, H. Handbuch der Gewerbehygiene auf experimenteller Grundlage. Berlin, 1876.
Fabrikärzte der deutschen chemischen Industrie. "Aerztliche Merkblätter über berufliche Vergiftungen." *Schriften aus dem Gesammtgebiet der Gewerbehygiene,* No. 1 (1913), new ser., No. 1 (1925), No. 28 (1930). Berlin.
—— "Was muss der Arzt von der neuen Verordnung über die Einbeziehung der Berufskrankheiten in die Unfallversicherung wissen und welche Pflichten ergeben sich für ihn daraus?" *Schriften aus dem Gesammtgebiet der Gewerbehygiene,* new ser., No. 14 (Berlin, 1925).
Fahr. "Ein Fall von Asbestosis." *Münchner Medicinische Wochenschrift,* LXI (1914), 625.
Fay, A. H. Coal Mine Fatalities in the U.S., 1870–1914. U.S. Bureau of Mines Bulletin 115. Washington, 1916.
Fieldner, A. G., S. H. Katz, H. W. Frevert, and E. G. Meiter. Gasmasks for Protection in Air against All Gases, Vapors and Smokes. U.S. Bureau of Mines, Report 2719. Washington, 1925.

Finlaison, A. G. Return on Sickness and Mortality Experiences of Friendly Societies. London, 1853, 1854.

Fischer, Alfons. Ein sozialhygienischer Gesetzentwurf aus dem Jahre 1800; ein Vorbild für die Gegenwart. Berlin, 1913.

—— Geschichte des deutschen Gesundheitswesens. Berlin, 1933.

Fischer, I. Geschichte des Gewerbedermatosen. In Vol. I of *Die Schädigungen der Haut durch Beruf und gewerbliche Arbeit*, ed. M. Oppenheim, J. A. Rille, and K. Ullmann. Leipzig, 1922.

Fischer, R. Die industrielle Herstellung und Verwendung der Chromverbindungen und die dabei entstehenden Gesundheitsgefahren. Berlin, 1911.

Flinn, R. H., J. L. Jones, and Others. Soft Coal Miners; Health and Working Environment. Public Health Bulletin 270. Washington, 1941.

Flury, F. "Blei." In *Handbuch der experimentellen Pharmakologie*, ed. by Heffter and Heubner. Berlin, 1934.

Flury, F., and H. Zangger. Lehrbuch der Toxikologie. Berlin, 1928.

Flury, F., and F. Zernik. Schädliche Gase. Berlin, 1931.

Forbes, J. J., M. J. Ankeny, and F. Feehan. Coal Miners' Safety Manual; a Handbook for Miners. U.S. Bureau of Mines, Washington, 1942.

Ford, C. B., and A. C. Stern. "Occupational Hazards in Fabrication of Magnesium and Its Alloys." New York State Dept. of Labor, Bulletin 23 (1944).

Fourcroy, A. F. de. Essai sur les maladies des artisans, traduit du Latin de Ramazzini. Paris, 1777.

Frank, H. " 'Caisson-krankheit' des Hüftgelenkes." *Münchner Medizinische Wochenschrift*, LXXXII (1935), 475.

Frank, Johann Peter. System einer vollständigen medizinischen Polizei. 6 vols., 1776–1817; 2 Supplements. Mannheim, Tübingen, Vienna, 1812, 1825.

Frankfurter Krankheitstafeln. Untersuchungen über Erkrankungsgefahr und Erkrankungshäufigkeit . . . auf Grund des Materiales der Ortskrankenkassen zu Frankfurt-am-Main. Elaborated by H. Bleicher. Frankfort-on-Main, 1900.

Friedrich, W. Die Phosphornekrose in Ungarn. Jena, 1910.

BIBLIOGRAPHY

Füller. "Hygiene der Berg- und Tunnelarbeiter." In *Gewerbehygiene:* Vol. VIII of *Handbuch der Hugiene*, ed. by T. Weyl. Jena, 1897.

Gafafer, William, ed. Manual of Industrial Hygiene and Medical Services in War Industries. Philadelphia, 1943.

Gardner, Leroy U. "Studies on the Relation of Mineral Dusts to Tuberculosis." *American Review of Tuberculosis*, Vols. IV (1920), VI (1922), VII (1923), XX (1929).

—— "Studies on Experimental Pneumokoniosis." *Journal of Industrial Hygiene*, XIV (1932), 18–37.

Gardner, Leroy U., and D. S. McCrum. "Effects of Daily Exposures to Arc-welding Fumes upon Animals." *Journal of Industrial Hygiene and Toxicology*, XXIV (1942), 173–182.

Geissler, A. Die Sterblichkeit und Lebensdauer der sächsischen Aerzte. Leipzig, 1887.

Gerlitt, John. "The Development of Quarantine." *Ciba Symposia*, II (1940), 566–580.

Glibert, D. Le Travail industriel des peaux, des poils et des crins. Brussells, 1921.

Goldwater, L. I. "From Hippocrates to Ramazzini: Early History of Industrial Medicine." *Annals of Medical History*, VIII (1936), 27–35.

Graefe, Alfred "Nystagmus der Bergleute." *Deutsche Medizinische Wochenschrift*, II, 260–261.

Grandhomme. Die Theerfarben-Fabriken der A. G. Farbwerke vorm. Meister Lucius und Brüning. Heidelberg, 1883.

Greenburg, L. S., H. Katz, G. W. Smith, and Others. Comparative Tests of Instruments for Determining Atmospheric Dusts. Public Health Bulletin 144. Washington, 1925.

Greenburg, L., M. R. Mayers, L. J. Goldwater, and A. Ross-Smith. "Benzol Poisoning in the Rotogravure Printing Industry in New York City." *Journal of Industrial Hygiene and Toxicology*, XXI (1939), 395–420.

Greenburg, L., and S. Moskowitz. "Occupational Diseases and Hazards in the Chemic and Rubber Industry." New York State Department of Labor, *Monthly Review* (November and December, 1946).

Greenburg, L., M. R. Mayers, and A. Ross-Smith. "Systemic Effects from Exposure to Certain Chlorinated Hydrocarbons." New York State Dept. of Labor, Bulletin (1939); also in *Journal of Industrial Hygiene and Toxicology*, XXI (1939), 29-38.

Greenburg, L., A. Ross-Smith, and J. Seigel. "Respiratory Diseases among Grain Handlers." New York State Department of Labor, *Industrial Bulletin*, 1941.

Greenburg, L., and W. Smith. A New Instrument for Sampling Aerial Dust. U.S. Bureau of Mines, Serial No. 2392. Washington, 1922.

Greenburg, L., and C. E. A. Winslow. "The Dust Hazard in Air Pressure Abrasive Blasting (Sandblasting)." *Archiv für Gewerbepathologie*, III (1932), 577-599.

Greenhow, H. H. "Third Series of Cases Illustrating the Pathology of Pulmonary Disease." In *Pathological Transactions*, London, 1861, 1869.

Greenwood, M. "Occupational and Economic Factors of Mortality." *British Medical Journal* (1939), pp. 862-866.

—— "The Evolution of an Industrial Society." *British Journal of Industrial Medicine*, I (1944), 2-6.

Gregory, I. C. "Case of Peculiar Black Infiltration of the Whole Lungs, Resembling Melanosis." *Edinburgh Medical and Surgical Journal*, XXXVI (1831), 389.

Grotjahn, A., and I. Kaup, eds. Handwörterbuch der sozialen Hygiene. Leipzig, 1912.

Grove, G. W. Self-contained Oxygen Breathing Apparatus; a Handbook for Miners. U.S. Bureau of Mines, Washington, 1941.

Guy, W. A. "On the Duration of Life of the Members of Several Professions." *Journal of the Statistical Society*, Vol. IX (1846).

Haldane, J. S. "Health and Safety in British Coal Mines." Chapter XVIII in *Historical Review of Coal Mining*. London, 1924.

Halfort, A. C. L. Entstehung, Verlauf und Behandlung der Krankheiten der Künstler und Handwerker. Berlin, 1845.

Hamilton, Alice. "White-Lead Industry in the United States, with an Appendix on the Lead-Oxide Industry." U.S. Bureau of Labor, Bulletin 95. Washington, 1911.

—— Hygiene of the Painters' Trade. U.S. Bureau of Labor Statistics, Bulletin 120 (Industrial Accidents and Hygiene and Accidents Series No. 2). Washington, 1912.

—— Lead Poisoning in the Smelting and Refining of Lead. U.S. Bureau of Labor Statistics, Bulletin 141. Washington, 1914.

—— Industrial Poisons Used or Produced in the Manufacture of Explosives. U.S. Bureau of Labor Statistics, Bulletin 219. Washington, 1917.

—— Industrial Poisonings in Making Coal Tar Dyes and Dye Intermediates. U.S. Bureau of Labor Statistics, Bulletin 280. Washington, 1921.

—— Industrial Poisons in the United States. New York, 1925.

—— Industrial Toxicology. New York, 1934.

—— Occupational Poisonings in the Viscose Rayon Industry. U.S. Bureau of Labor Statistics, Bulletin 34. Washington, 1940.

—— Exploring the Dangerous Trades. New York, 1943.

Hamilton, Alice and Others. Effect of Airhammer on the Hands of Stonecutters. U.S. Bureau of Labor Statistics, Bulletin 236, 1918.

Hamilton, Alice, and R. T. Johnstone. "Industrial Toxicology." *Oxford Medicine*, Vol. IV, No. 2 (London, 1927). Reprinted, 1947.

Handbuch der Hygiene, ed. by T. Weyl. Vol. VII: *Gewerbehygiene*. 1st ed., Jena, 1897; 2d ed., Jena, 1913.

Handbuch der Sozialen Hygiene und Gesundheitsfürsorge, ed. by A. Gottstein, A. Schlossmann, and L. Teleky. Vol. II: *Gewerbekrankheiten und Gewerbehygiene*. Berlin, 1926.

Hannover, A. Die Krankheiten der Arbeiter. Supplement to *Deutschen Klinik*, 1861.

Harvey, V. K., and E. P. Luongo. "Physical Capacity for Work." *Occupational Medicine*, II (1946), 1–47.

Hatlapa, W. "Umriss zur hundertjährigen Geschichte der deutschen Arbeitsaufsicht." *Archiv für Gewerbepathologie*, VI (1935), 222–256.

Hatzfeld, K. "Die Mittel zur Bekämpfung von Grubenexplosionen in England und Frankreich und ihre Anwendung im Deutschen Steinkohlenbergbau." *Zeitschrift für das Berg-,*

Hütten- und Salinenwesen im preussischen Staate. LXVI, 1918.

—— "Die Entwicklung der Massnahmen zur Kohlenstaub bekämpfung." *Berg- und Hüttenmännische Zeitschrift "Glückauf."* 1925.

Hebenstreit, E. B. G. Lehrsätze der medizinischen Polizeiwissenschaft. Leipzig, 1791; 2d ed., Vienna, 1806.

Hebestreit, H. "Der Schutz der menschlichen Areitskraft im Rahmen der Sozialpolitschen Neuordnung." *Zentralblatt für Gewerbehygiene,* new ser., XIII (1936), 49–53.

Hedley, O. F. "Medical Services." Chapter IV in *Manual of Industrial Hygiene,* ed. by W. Gafafer. Philadelphia, 1943.

Heijermans, L. Handleiding tot de Kennis der Beroepsziekten. Rotterdam, 1908.

Heiman, H., and C. B. Ford. "Chronic Benzol Poisoning in the Plastic Industry." New York State Dept. of Labor, Bulletin, 1940.

Heinzerling, C. Die Gefahren und Krankheiten der chemischen Industrie und die Mittel zu ihrer Verhütung und Beseitigung. Halle, 1886–1887.

Heller, R., W. Mager, and H. von Schrötter. Luftdruckerkrankungen. 2 vols. Vienna, 1900.

Henckel, Johann Friedrich. Medizinischer Aufstand und Schmelzbogen, von der Bergsucht und Hüttenkatze; auch einige andern den Bergleuten und Hüttenarbeitern zustossenden Krankeiten. Dresden and Leipzig, 1745.

Henderson, Y., and H. W. Haggard. Noxious Gases. 1st ed., New York, 1927; 2d ed., New York, 1943.

Heymann, B., and F. Freudenberg. Morbidität und Mortalität der Bergleute im Ruhrgebiet. Essen, 1925.

Hill, Leonard. Caisson Sickness. London, 1912.

—— The Science of Ventilation and Open Air Treatment. Parts I and II. London, 1919, 1920.

Hirt, Ludwig. Die Krankheiten der Arbeiter. 4 vols. Breslau and Leipzig, 1871–1878.

Hirt, Ludwig, and G. Merkel. Die Gewerbekrankheiten. In *Handbuch der Hygiene und Gewerbekrankheiten,* ed. by M. von Pettenkofer and H. von Ziemssen. Leipzig, 1882.

Historical Review of Coal Mining. London, 1924.
Hoffman, Frederic L. "The Mortality from Consumption in Dusty Trades." U.S. Bureau of Labor, Bulletin 17 (Washington, 1908–9), pp. 635–875.
—— The Mortality from Respiratory Diseases in Dusty Trades (Inorganic Dust). Bureau of Labor Statistics, Bulletin 231. Washington, 1918.
Hoffmann, Friedrich. De metallurgia morbifica. Halle, 1705.
—— Eines berühmten Medici gründliches Bedencken und physikalische Anmerkungen von den tödlichen Dampf der Holtzkohlen. Halle, 1716. Reprinted in *Voigtländers Quellenbücher*, Vol. XIV (1912).
Holland, G. C. The Vital Statistics of Sheffield. London, 1843.
Holland, J. W. A Textbook of Medical Chemistry and Toxicology. Philadelphia, 1906.
Hope, E. W., W. Hanna, and C. O. Stallybrass. Industrial Hygiene and Medicine. London, 1923.
Houghten, F. C., C. F. Yagloglou, and R. R. Sayers. Effective Temperature for Still-Air Conditions and Their Application to Mining. Bureau of Mines, Report 2563. Washington, 1924.
Hue, O. Die Bergarbeiter. Stuttgart, 1910–1913.
Hyde, R. "Medical Services in Industry in Great Britain." *International Labor Review*, LI (1945), 433–438.
Ilsley, L. C., and A. B. Hooker. Permissible Electric Mine Lamps. U.S. Bureau of Mines, Bulletin, 332 (1930) and 441 (1942).
Ins, Adolf von. Experimentelle Studien über Kieselstaubinhalation. *Archiv für experimentelle Pathologie*, V (1876), 169–194.
Institut für Gewerbehygiene zu Frankfort-am-Main. Mitteilugen des . . . Beiblatt zur Sozialtechnik. Berlin, 1910–1912.
Institute of Civil Engineers. Report of the Committee on Regulations for Guidance of Engineers and Contractors for Work Carried Out under Compressed Air. London, 1936.
International Association of Governmental Labor Officials, 26th Convention. Labor Laws and Their Administration. U.S. Dept. of Labor, Bureau of Statistics, Bulletin 690. Washington, 1941.

Internationale Hygiene Ausstellung, Dresden, 1911; Historische Abteilung, Catalogue.
International Labor Office. Health and Occupation. Geneva, 1926-1933. (Supplements, to 1944, Montreal.)
—— Cancer of the Bladder among Workers in Aniline Factories. 1921.
—— Compensation for Industrial Accidents 1925. (Comparative Analysis of National Laws.)
—— Compensation for Occupation Diseases 1925. (Comparative Analysis of National Laws.)
—— Factory Inspection. Historical Development and Present Organisation in Certain Countries. 1925.
—— Medical Inspection of Labour (Report of the Meeting of Medical Inspectors of Labour held in Dusseldorf 1926.)
—— White Lead. 1927. (Data collected by the I.L.O. in regard to the use of white lead in the painting industry.)
—— Statistical Methods for Measuring Occupational Morbidity and Mortality. 1930.
—— Acetylene: Safety in the manufacture and use of. 1931.
—— Celluloid: Safety in the Manufacture and Use of. 1933.
—— Safety in Spray Painting. 1938.
Jacobs, M. B. The Analytical Chemistry of Industrial Poisons, Hazards and Solvents. New York, 1941.
Jaeger, F. "Caissonkrankheit des Hüftgelenkes." *Monastsschrift für Unfallheilkunde*, XLIV (1937), 374-382.
Jaquet, A. "Ueber Brommethylvergiftung." *Deutscher Archiv für Klinische Medizin*, LXXI (1901), 370-386.
Jessop. "An Extract of a Letter of July 28, 1675, Communicated by Mr. Lister . . . Containing Some Observations about Damps in Mines." *Philosophical Transactions*, X (1675), No. 117.
Jones, D. C. Coal Mining. Philadelphia (?), 1941.
Jones, R. R., J. W. Crosson, R. R. Sayers, H. H. Schrenk, and E. Levy. "Administration of Pure Oxygen to Compressed-Air Workers during Decompression." *Journal of Industrial Hygiene and Toxicology*, XX (1940), 427-444.
Jones, W. R. "The Silicotic Lungs: the Minerals They Contain." *Journal of Hygiene*, XXX (1933), 307-329.

Johnstone, T. Rutherford. Occupational Diseases. New York, 1941.
Kalstrom, S. C., C. C. Burton, and D. B. Phemister. "Infraction of Bones in Caisson Diseases." *Surgery, Gynaecology and Obstetrics*, LXI (1939), 129–146.
Karpeles, Benno. Die englischen Fabriksgesetze. Berlin, 1900.
Karup, Gollmer, and Florschütz. Aus der Praxis der Gothaer Lebensversicherungsbank. Jena, 1902.
Katz, S. H., G. W. Smith, and E. G. Meiter. Dust Respirators, Their Construction and Filtering Efficiency. U.S. Bureau of Mines, Technical Paper 394. Washington, 1926.
Katz, S. H., G. W. Smith, W. M. Myers, L. J. Trostel, M. Ingals, and L. Greenburg. Comparative Tests of Instruments for Determining Atmospheric Dusts. U.S. Public Health Bulletin 144. Washington, 1925.
Keays, F. L. Compressed Air Illness, a Report of 3,692 Cases. New York, 1909.
Kircher, Athanasius. Mundus subterraneus. Amsterdam, 1665.
Klein, G. A. Statistik der Arbeiterversicherug des Deutschen Reichs für die Jahre 1885–1906. Berlin, 1908.
Kober, G. M. "History of Industrial Hygiene," In *A Half-Century of Public Health*, ed. by M. P. Ravenal. New York, 1921.
—— "Historical Review of Industrial Hygiene and Its Effects on Public Health." In *Industrial Health*, ed. by G. M. Kober and E. Hayhurst. Baltimore, 1924.
Kober, G. M., and W. C. Hanson, eds. Diseases of Occupation and Vocational Hygiene. Philadelphia, 1916.
Kober, G. M. and E. R. Hayhurst. Industrial Health. Baltimore, 1924.
Kobert, R. Lehrbuch der Intoxicationen. 1st ed., Stuttgart, 1893; 2d ed., Stuttgart, 1902–1906.
Koelsch, F. "Gesundheitliche Erhebungen über das Malerbewerbe." In *Erhebungen der Bayerischen Gewerbeaufsichtsbeamten über das Maler gewerbe*. Munich, 1912.
—— "Staub und Beruf." In *Handwörterbuch der Sozialen Hygiene*, ed. by A. Grotjahn and I. Kaup. Leipzig, 1912.

Koelsch, F. "Die Giftwirkung des Zyanamids." *Zentralblatt für Gewerbehygiene* (1916), 113–120.

—— Die Bleischädigungen im Maler- und Lackierergewerbe. Hamburg, 1921.

—— "Theophrastus von Hohenheim genannt Paracelsus, 'Von der Bergsucht und andern Bergkrankheiten.' " *Schriften aus dem Gesammtgebiet der Gewerbehygiene*, new ser., No. 12 (Berlin, 1925).

—— "Die Gesundheitsverhältnisse der Arbeiter in Lumpensortir-Betrieben." *Reichsarbeitsblatt* (1930), Part III: "Arbeitsschutz," pp. 116–119.

—— "Untersuchungen über Staubgefährdung in Chamottefabriken." *Reichsarbeitsblatt* (1932), Part III: "Arbeitsschutz," pp. 2–16.

—— Handbuch der Berufskrankheiten. Jena, 1935, 1937.

—— Lehrbuch der Arbeitshygiene. Stuttgart, 1946–1947.

Kossoris, M. D., and S. Kjaer. "Industrial Injuries in the U.S. during 1939." *Monthly Labor Review*, Vol. LI (1940), *et seq*.

Kraft, M. Fabrikhygiene. Vienna, 1891.

Krankheitsstatistik der Allgemeinen Ortskrankenkasse der Stadt Berlin, 1915–1918. Berlin, 1921.

Krüger, Elisabeth, Rostoski and Saupe. "Ueber Lungenasbestose." *Archiv für Gewerbepathologie*, II (1931), 558–590.

Kunkel, A. J. Handbuch der Toxikologie. Jena, 1899–1901.

Kussmaul, A. Untersuchungen über den constitutionellen Merkurialismus und sein Verhältnis zur constitutionellen Syphilis. Würzburg, 1861.

Lanza, A. J., and S. B. Childs. Miners' Consumption. Public Health Bulletin 85. Washington, 1917.

Lanza, A. J., and E. Higgins. Pulmonary Diseases among Miners in the Joplin District. U.S. Bureau of Mines, Technical Paper 105. Washington, 1915.

Lanza, A. J., and J. A. Goldberg, eds. Industrial Hygiene. New York, 1939.

Layet, A. Hygiene des professions et des industries. Paris, 1875.

Legge, T. M. In the *Annual Report of the Chief Inspector of Factories, 1900* (London, 1901), Appendix I (on ganister mines).

——— *The Health of Brassworkers.* In *Report of the Chief Inspector of Factories, 1905* (London, 1906), Appendix I.

——— Ulcerations of the Skin and Epitheliomatous Cancer in the Manufacture of Patent Fuel and of Grease. London, 1910.

——— "Report on Body Temperature of Spinners and Weavers." Appendix IX to the *Report of the Departmental Committee on Humidity in Flax Mills and Linen Factories.* London, 1914.

Legge, T. M., A. M. Anderson, and G. E. Duckering. Special Report on Dangerous and Injurious Processes in the Coating of Metal with Lead. London, 1907.

Legge, T. M., and K. W. Goadby. Lead Poisoning and Lead Absorption. London, 1912.

——— Bleivergiftung und Bleiaufnahme. Tr. by H. Katz, with annotations by L. Teleky. Berlin, 1921.

Lehmann, K. B. Kurzes Lehrbuch der Arbeits- und Gewerbehygiene. Leipzig, 1919.

Lehmann, K. B., and Others. "Experimentelle Studien über den Einfluss technisch wichtiger Gase und Dämpfe auf den Organismus." *Archiv für Hygiene,* 1886, *et seq.*

Leigh, C. The Natural History of Lancashire, Cheshire, and Derbyshire. London, 1700.

Levy, E. Compressed-Air Illness and Its Engineering Importance, with a Report of Cases at the East River Tunnels. Bureau of Mines, Technical Paper 285. Washington, 1922.

Lewin, L. Gifte und Vergiftungen. 4th ed. of *Lehrbuch der Toxikologie* (1885). Berlin, 1929.

——— Die Kohlenoxydvergiftung. Berlin, 1920.

Leymann. Die Bekämpfung der Bleigefahr in der Industrie. Jena, 1908.

——— "Bleierkrankungen unter den bei der Herstellung keramischer Abziehbilder beschäftigten Arbeitern." *Zentralblatt für Gewerbehygiene,* II (1914), 223–225.

——— "Ein Beitrag zur Geschichte des technischen Gefahrenschutzes in Deutschland." *Zentralblatt für Gewerbehygiene,* new ser., XII (1935), 2–7, 49–54.

Lincoln, D. F. School and Industrial Hygiene. Philadelphia, 1880.

Linden, Mrs. Bates J. W. Mercury Poisoning in the Industries of New York City and Vicinity. New York, 1912.

Linné, Karl von. Iter Dalecarlicum. 1734.

Lipkowič, J. "Zur Frage des Einflusses von Erdboden und Menge der zugeführ ten Luft auf die Haüfigkeit der Casissonkrankheit." *Archiv für Gewerbepathologie*, VII (1937), 378–382, 594.

Lipson, E. The Economic History of England. London, 1937.

Littlefield, J. B., F. L. Feicht, and H. H. Schrenk. Bureau of Mines Midget Impinger for Dust Sampling. U.S. Bureau of Mines, Report 3360. Washington, 1937.

Löbker, and H. Bruns. "Ueber das Wesen und die Ausbreitung der Wurmkrankheit (Ankylostomiasis) mit besonderer Berücksichtigung ihres Auftretens im deutschen Bergbau." *Arbeiten aus dem Kaiserlichen Gesundheitsamt*, XXIII (1906), 421–524.

Löhneiss, G. E. Bericht von Bergwercken. Hamburg and Stockholm, 1690.

Lombard, H. C. De l'influence des professions sur la durée de la vie. Geneva, 1835.

Lorinser. "Jahresbericht über die chirurgische Abteilung des Bezirkskrankenhauses Wieden in Wien, 1843: Nekrose der Kieferknochen infolge Einwirkung von Phosphordämpfen." *Medizinische Jahrbücher des kaiser-königlichen oesterreichischen Staates* (1845), p. 257.

Lubenau, C. "Experimentelle Staubinhalationskrankheiten der Lunge." *Archiv für Hygiene*, LXIII (1907), 391–409.

McCready, Benjamin W. The Influence of Trades, Professions and Occupations in the United States on the Production of Diseases. New York, 1937; new ed., with an Introductory Essay by Geneviève Miller, Baltimore, 1943.

McNally, W. D. Toxicology. Chicago, 1937.

Mai, Franz Anton (Stolpertus). Entwurf einer Gesetzgebung über die wichtigsten Gegenstände der medizinischen Polizei als Beitrag zu einem neuen Landrecht in der Pfalz. Mannheim, 1802.

Maler Deutschlands, Vorstand der Vereinigung der. Der Kampf gegen die giftigen Bleifarben. Hamburg, 1904.

Mallette, F. S. "The Role of Industrial Hygiene in Labor Relations." *Industrial Medicine*, XIII (1944), 775–777.
Mayer, R. L. Das Gewerbeeczem. Berlin, 1930.
Mayers, M. R. Carbon Monoxide Poisoning in Industry and Its Prevention. New York State Department of Labor, Special Bulletin. Albany, N.Y., 1930.
Mayers, M. R., and M. McMahon. Lead in Industry. New York State Dept. of Labor, Special Bulletin 195. Albany, N.Y., 1938.
Meinel, F. Ueber die Erkrankungen der Lungen durch Kieselstaubinhalation. Erlangen, 1869.
Meissner. "Hygiene der Berg- und Tunnelarbeiter." In *Handbuch der Hygiene*, ed. T. Weyl, Vol. VIII: *Gewerbehygiene*. Jena, 1895.
Mérat, F. V. Traité de la colique metallique. 1st ed., Paris, 1803; 2d ed., Paris, 1812.
Merewether, E. R. A. "The Occurrence of Pulmonary Fibrosis and Other Pulmonary Infections in Asbestos Workers." *Journal of Industrial Hygiene*, XII (1930), 198–222, 239–257.
Merewether, E. R. A., and C. W. Price. Report on the Effects of Asbestos Dust on the Lungs and Dust Suppression in the Asbestos Industry. London, 1930.
Middleton, E. L. "Industrial Pulmonary Disease Due to the Inhalation of Dust." *Lancet*, II (1936), 59–64.
Middleton, E. L., and E. L. Macklin. Report on the Grinding of Metals and Cleaning of Castings. London, 1923.
Mining Association of Great Britain. Historical Review of Coal Mining. London, 1924.
Morrah, D. A Historical Outline of Coalmining Legislation. Chapter XX in *Historical Review of Coal Mining*. London, 1924.
Morris, G. E., I. R. Tabershaw, and Others. "Protection of Radium Dial Painters: Specific Work Habits and Equipment." *Journal of Industrial Hygiene and Toxicology*, XXV (1943), 270–274.
Moskowitz, S., and W. J. Burke. "Control of Health Hazards from Rubber Cementing Operations in Shoe and Slipper Manufacture." New York State Dept. of Labor, *Monthly Review* (January–February, 1946).

Moss, K. N. "Mine Gases and Explosions." Chapter IX in *Historical Review of Coal Mining*. London, 1924.

Müller, Richard. Die Bekämpfung der Bleigefahr in Bleihütten. Jena, 1908.

National Conference on Labor Legislation, Twelfth, December, 1945. Résumé of the Proceedings. U.S. Dept. of Labor, Division of Labor Standards, Bulletin 76. Washington, 1946.

National Safety Council, Chicago, Ill. Annual publications: *Accident Facts; National Safety News*, 1919–; *Public Safety*, 1927–.

Neal, P. A., R. H. Flinn, and Others. Chronic Manganese Poisoning in an Ore-Crushing Mill. Public Health Bulletin 24. Washington, 1910.

Neal, P. A., R. R. Jones, J. J. Bloomfield, and Others. Mercurialism in the Hat and Fur-Cutting Industry. Public Health Bulletins 234 and 263. Washington, 1937, 1941.

Nef, J. U. The Rise of the British Coal Industry. London, 1931.

Neison, F. G. P. Conditions of Vital Statistics. 3d ed. London, 1857.

Neisser, E. J. Internationale Uebersicht über Gewerbehygiene nach den Berichten der Gewerbeinspectionen der Kulturländer. Berlin, 1907.

Nelson, W. T. Report on an Investigation of the Pulmonary Conditions of Mine Employees of Western Australia, during 1925–1926. Division of Industrial Hygiene, Service Publication No. 5. Canberra, Australia.

Neufville, W. C., de. Lebensdauer und Todesursachen 22 verschiedener Stände und Gewerbe. Frankfort-on-Main, 1855.

New York State Factory Investigation Commission. Preliminary Report. Albany, N.Y., 1912.

—— Second, Third, and Fourth Reports. Albany, 1913, 1914, and 1915.

Nicloux, M. L'Oxyde de carbone et l'intoxication oxycarbonique. Paris, 1925.

Nieden, A. "Ueber Nystagmus als Folgezustand von Hemeralopie." *Berlin er klinische Wochschrift* (1874), 593–596.

—— Der Nystagmus der Bergleute. Wiesbaden, 1894.

Norris, E. H. "A School of Industrial Health." *Industrial Medicine*, XIII (1944), 780–782.
Oesterlen, F. Handbuch der medizinischen Statistik. 1st ed. Tübingen, 1865; 2d ed., Tübingen, 1874.
Oettingen, W. F., von. The Aromatic Amino- and Nitrocompounds. Public Health Bulletin 271. Washington, 1941.
—— The Aliphatic Alcohols. Public Health Bulletin 281. Washington, 1943.
—— Carbon Monoxide, Its Hazards. Public Health Bulletin 290. Washington, 1944.
Oettingen, W. F., von, D. D. Donahue, and R. K. Snyder. Trinitrotoluene. Public Health Bulletin 285. Washington, 1944.
Ohm, J. "Akkomodationskrampf und Augenzittern der Bergleute." *Klinische Monatsblatter für Augenheilkunde*, XLVIII (1910), 608–611.
—— Das Augenzittern der Bergleute und Verwandtes. Berlin, 1916.
Oldendorff, A. Der Einfluss der Beschäftigung auf die Lebensdauer der Menschen. Berlin, 1878.
Oliphant. Report on Friendly or Benefit Societies Exhibiting the Law of Sickness as Deduced from Returns by Friendly Societies in Different Parts of Scotland. Edinburgh, 1824.
Oliver, Sir Thomas, ed. Dangerous Trades. London, 1902.
—— "Industrial Lead Poisoning with Description of Lead Processes in Certain Industries in Great Britain." Bureau of Labour, Bulletin No. 95. Washington, 1911.
Orfila, R. M. Traité des poisons. 4 vols. Paris, 1814–1815.
Owens, J. S. "Jet Dust-counting Apparatus." *Journal of Industrial Hygiene*, IV (1922–23), 522–534.
Pansa Martin. Consilium peripneumoniacum: Ein getreuer Rat in der beschwerlichen Berg- und Lungensucht. Leipzig, 1614.
Paracelsus (Theophrastus Bombast von Hohenheim genannt Paracelsus). Von der Bergsucht und anderen Bergkrankheiten. 3 vols. Dillingen, 1567; ed. by F. Koelsch, in *Schriften aus dem Gesammtgebiet der Gewerbekrankheiten*, new ser., No. 12 (Berlin, 1925).
Paraf, Georges G. Hygiène et sécurité du travail industriel. Paris, 1905.

Parson, Allan C. "Abstract of Report on Bakers' Itch." *Journal of Industrial Hygiene*, V (1923-24), 410-433.
Patissier, P. Traité des maladies des artisans d'après Ramazzini. Paris, 1822.
Patschkowski. "Ueber Pneumonokoniosen bei den Bergarbeitern des rheinischwestfälischen Steinkohlenreviers." *Beiträge zur Klinik der Tuberkulose*, LVII (1923), 113.
Pearson, G. In *Philosophical Transactions*, Part II (1813), p. 159.
Petri, Else. Pathologische Anatomie und Histologie der Vergiftungen. Vol. X of the *Handbuch der speciellen pathologischen Anatomie und Histologie*. Berlin, 1930.
Pieraccini, G. Patologia del lavoro e terapia sociale. Milan, 1906.
Pohlen, K. Gesundheitsstatistisches Auskunftsbuch für das Deutsche Reich. Veröffentlichungen aus dem Gebiet der Medizinalverwaltung, No. 409. Berlin, 1936.
Porro, F. W., J. R. Patton, and A. A. Hobbe. "Pneumoconiosis in the Talc Industry." *American Journal of Roentgenology and Radiotherapy*, XLVII (1942), 507-529.
Potton, F. A. F. "Recherches et observations sur le mal de vers ou mal de bassine." *Bulletin de l'Academie imperial de médicine*, Vol. XVII (1851-52).
Pratt, E. E. "Occupational Diseases: Preliminary Report on Lead Poisoning in the City of New York, with an Appendix on Arsenical Poisoning." New York State Factory Investigating Commission, Preliminary Report, Appendix VI. Albany, N.Y., 1912.
Preti, Luigi. Trattato di patologia medica del lavoro. Milan, 1941.
Price, George M. A General Survey of the Sanitary Conditions of the Shops in the Cloak Industry. First Annual Report of the Joint Board of Sanitary Control in the Cloak, Suit and Shirt Industry of Greater New York. New York, 1911.
—— The Modern Factory. New York, 1914.
Prinzing, F. Handbuch der medizinischen Statistik. 1st ed., Jena, 1906; 2d ed., Jena, 1930-1937.
Pütsch, A. Die Sicherung der Arbeiter gegen die Gefahren für Leben und Gesundheit im Fabrikbetriebe. Berlin, 1883.

Ramazzini, Bernardo. De morbis artificum diatriba. Modena, 1700; also in Bernardini Ramazzini, *Opera Omnia*, Geneva, 1717.
Raneletti, Aristide. Le Malattie del lavoro. Rome, 1924.
Raymond, R. W. "Hygiene of Metal Mines." In *Cyclopaedia of the Practice of Medicine*, ed. by H. von Ziemssen. New York, 1879.
Rehn, L. "Blasengeschwülste bei Fuchsinarbeitern." *Archiv für klinische Chirurgie*, L (1895), 588–600.
Reichel, F. Die Sicherung von Leben und Gesundheit im Fabrik- und Gewerbebetrieb auf der Brüsseler Ausstellung, 1876. Berlin, 1877.
Reynal. "De l'herpes tonsurant dans les espèces chevaline et bovine contagieux." *Bulletin de l'Acadamie imperial de médecine*, XXIII (1857), 223.
Rice, George S., and J. Hartmann. "Coal Mining in Europe. U.S. Bureau of Mines, Bulletin 414. Washington, 1939.
Rice, George S., and Others. Coaldust Explosions Tests in Experimental Mines, 1913–1932. U.S. Bureau of Mines, Bulletins 56, 167, 268, 369.
Riley, Cassius M. Toxicology. St. Louis, Mo., 1902.
Rode, C. D. "Ueber den Nystagmus und seine Ursachen." Dissertation, Halle, 1874.
Roger. In *Philosophical Transactions*, 1676, 1677.
Rosen, George. The History of Miners' Diseases. New York, 1943.
Rosenfeld, Siegfried. "Die Gesundheitsverhältnisse der Wiener Arbeiterschaft." *Statistische Monatsschrift*, 1905–6.
Ross, A. A., and N. H. Shaw. Dust Hazards in Australian Foundries. Commonwealth of Australia, Dept. of Labour, Technical Report No. 1. Melbourne, 1943.
Royal College of Physicians. Social and Preventive Medicine Committee. "Report." *British Journal of Industrial Medicine*, II (1945), 51–55.
Royal Society of Medicine. The Origin, Symptoms, Pathology, Treatment, and Prophylaxis of Toxic Jaundice Observed in Munitions Workers. London, 1917.
Sachs, Otto. Gewerbekrankheiten der Haut. In Vol. I, *Hand-*

buch der Haut und Geschlechtskrankheiten, ed. by J. Jadassohn. Berlin, 1930.

Sachs, Willy. Die Kohlenoxydvergiftung. Braunschweig, 1900.

Sayers, R. R., J. J. Bloomfield, J. M. Dallavalle, and Others. Anthraco-silicosis among Hard Coal Miners. Public Health Bulletin 221. Washington, 1936.

Schablowski. "Ueber Respiratoren bei gewerblicher Staubarbeit." *Zeitschrift für Hygiene und Infectionskrankheit,* LXVIII (1911), 169.

Scheffler, C. L. Abhandlung von der Gesundheit der Bergleute. Chemnitz, 1770.

Schlattmann, H. Sammlung der für den Oberbergamtsbezirk Dortmund geltenden wichtigsten bergpolizeilichen Verordnungen und Bestimmungen nebst Erläuterungen. 8th ed. Essen, 1929.

Schlockow, I. Die Gesundheitspflege und medizinische Statistik im preussischen Bergbau. Berlin, 1881.

Schneider, H. Gefahren der Arbeit in der chemischen Industrie. Hannover, 1911.

Schoenlank, Bruno. Die Fürther Quecksilber-Spiegelbelegen und ihre Arbeiter. Stuttgart, 1888.

Schrenk, H. H. Testing and Design of Respiratory Protective Devices. Bureau of Mines, Information Circular 7086. Washington, 1939.

Schrenk, H. H., and W. D. Foster. Petrographic Identification of Atmospheric Dust Particles. Bureau of Mines, Report 3368. Washington, 1938.

Schrenk, H. H., S. J. Pearce, and W. P. Yant. A Microcolorimetric Method for the Determination of Benzene. Bureau of Mines, Report 3297. Washington, 1935.

Schwartz, Louis. Skin Hazards in American Industries. Parts I and II. Public Health Bulletins 215 (Washington, 1934), 229 (Washington, 1936).

Schwartz, Louis and L. Tulipan. Occupational Disease of the Skin. Philadelphia, 1939; 2d ed., with S. M. Peck, Philadelphia, 1947.

Seltmann. "Die Anthrakosis der Lungen bei den Kohlenbergarbeitern." *Deutsches Archiv für klinischen Medizin*, II (1867), 300–326.

Sendtner, J. "Ueber Lebensdauer und Todesursachen bei den Biergewerben." *Münchner Medizinische Abhandlungen*, No. 2 (Munich, 1891).

Sheafer, H. C. "Hygiene of Coal Mines." In *Cyclopaedia of the Practice of Medicine*, ed. by H. von Ziemssen. New York, 1879.

Siegal, W., A. Ross-Smith, and L. Greenburg. "The Dust Hazards in Tremolite Talc Mining, Including Roentgen-Findings in Talcworkers." *American Journal of Roentgenology and Radiotherapy*, LXIX (1943), 11–29.

Silicosis. Records of the International Conference Held at Johannesburg, August, 1930. International Labor Office, Geneva, 1930.

Silicosis. Proceedings of the International Conference Held in Geneva, 1938. International Labor Office, Geneva, 1940.

Singstad, Ole. "Industrial Operations in Compressed Air." *Journal of Industrial Hygiene*, XVIII (1936), 497–523.

—— "Engineering Problems Related to the Health of Workers in Compressed Air." *Proceedings of the Sixth Pacific Science Congress*, VI (1939).

Snell, E. H. Compressed Air Illness. London, 1896.

Sollmann, Torald H. A Manual of Pharmacology and Its Application to Therapeutics and Toxicology. 1st ed., Philadelphia, 1917; 6th ed., Philadelphia, 1942.

Sommerfeld, Theodor. Handbuch der Gewerbekrankheiten, Vol. I. Berlin, 1898.

Sonne, W. "Hygiene der keramischen Industrie." In *Handbuch der Hygiene*, ed. by T. Weyl, Vol. VIII: *Gewerbehygiene*. 1897.

Southam, A. H., and S. R. Wilson. "Cancer of the Scrotum." *British Medical Journal*, II (1922), 971–973.

Starkenstein, E., E. Rost, and J. Pohl. Toxikologie. Vienna, 1929.

Staub-Oetiker. "Die Pneumonokoniose der Metallschleifer." *Deutsches Archiv für klinische Medizin*, CXI (1916), 469–481.

Stern, A. C., J. Baliff, and others. Characteristics of Unit Dust

Collectors. New York State Department of Labor, *Monthly Review* (August-September, 1946).

Stern, B. J. Medicine in Industry. New York, 1946.

Stiller. "Gewerbewesen und Gewerbepolizei um 1750 in Preussen sowie die Anfänge des Arbeiterschutzes." *Reichsarbeitsblatt*, Part III: "Arbeiterschutz" (1926), p. 1.

Stockhausen, Samuel. De lithargyri fumo noxio morbifico, ejusque metallico frequentiori morbo, vulgo die Hüttenkatze, cum appendice de montano asthmate, metallicis familiari, vulgo die Bergsucht. Goslar, 1656.

Stratton, Thomas. "Cause of Anthracosis or Black Infiltration of the Whole Lungs." *Edinburgh Medical and Surgical Journal*, Vol. LXIX (1838).

Summons, W. Miners' Phthisis at Bendigo. Melbourne, Australia, 1907.

Süssmilch, Johann Peter. Die göttliche Ordnung in den Veränderungen des menschlichen Geschlechtes. Berlin, 1761, 1762.

Sutherland, C. L., and S. Bryson. Report on the Incidence of Silicosis in the Pottery Industry. London, 1926.

—— Report on the Occurrence of Silicosis among Sandstone Workers. London, 1929.

—— Report on an Inquiry . . . in the Slate Industry in the Gwyrfai District. London, 1930.

Tabershaw, Irving R. "Radium Dial Painting: Medical Status of Workers." *Journal of Industrial Hygiene and Toxicology*, XXVIII (1946), 212-216.

Tanquerel des Planches, L. Traité des maladies de plomb ou saturnines. Paris, 1839.

—— Die gesammten Bleikrankheiten. Tr. by S. Frankenberg. Leipzig, 1842.

Tardieu, A. Etude medicolégale et clinique sur l'empoisonnement. 1st ed., Paris, 1867; 2d ed., Paris, 1875.

Teleky, L. Die Phosphornekrose: Ihre Verbreitung in Oesterreich und deren Ursachen. Vienna, 1907.

—— Die gewerbliche Quecksilbervergiftung. Berlin, 1912.

—— "Aufgaben und Durchführung der Krankheitsstatistik der Krankenkassen." *Veröffentlichungen aus dem Gebiete der Medicinalverwaltung*, Vol. 173, Berlin, 1923.

—— "Die Krankheitsstatistik der rheinischen Krankenkasses, 1922–1926." *Reichsarbeitsblatt* (1929), Supplement.
—— "Die Krankheitsstatistik der nach dem 'Rheinischen Schema' arbeitenden Krankenkassen, 1922–1931." *Archiv für Gewerbepathologie*, V (1934), 764–809.
—— Vorlesungen über soziale Medizin. Jena, 1914.
Teleky, L., and J. Lochtkemper. "Studien über Staublunge." *Archiv für Gewerbepathologie*, III (1932), 418–470, 600–769.
Teleky, L., J. Lochtkemper, E. Rosenthal-Deussen, and Derdack. "Staubgefährdung und Staubschädigungen der Metallschleifer." *Arbeit und Gesundheit*, No. 9 (Berlin, 1928).
Teleky, L., and I. Weikert. "Die Wirkung der Fabriksarbeit der Frau auf die Mutterschaft." *Arbeit und Gesundheit*, No. 14 (Berlin, 1930).
Teleky, L., and E. Zitzke. "Untersuchungen über das Bäckerecezem und seine Ursachen." *Archiv für Gewerbepathologie*, III (1932), 58–152.
Thackrah, C. Turner. The Effect of Arts, Trades and Professions and of Civic States and Habits of Living on Health and Longevity. 2d ed. London, 1832.
Thompson, L. R., D. K. Brundage, A. E. Russell, J. J. Bloomfield, and H. Britton. The Health of Workers in Dusty Trades. Public Health Bulletins 176, 187, 208. Washington, 1928, 1929, 1933.
Thompson, W. G. The Occupational Diseases. New York, 1914.
Thomson, W. "The Black Expectoration and the Deposition of Black Matter in Lungs." *Medicochirurgical Transactions*, Vols. XX (1837), XXI (1838).
Tolman, W. H., and L. B. Kendall. Safety. New York, 1913.
Tracy, Roger S. "Hygiene of Occupations." In *Cyclopaedia of the Practice of Medicine*, ed. by H. von Ziemssen. New York, 1879.
Traube, Ludwig. "Ueber das Eindringen feiner Kohlenteilchen in das Innere des Respirationsapparates." *Deutsche Klinik*, XII (1860), 475–478, 487–490.
Ullmann, Karl, M. Oppenheim, and J. H. Rille. Dis Schädigungen der Haupt durch Beruf und gewerbliche Arbeit. 3 vols. Leipzig, 1922–1926.

United Automobile Workers–CIO. Health Institute. Excerpts from the Report of the Medical Director to the Board of Trustees, Jan. 1–June 20, 1946. Mimeographed. Detroit, Mich., 1946.

Ursinus, M. Leonardus. De morbis metallorum. Leipzig, 1652.

Vernois, Maxime. "De la main des ouvriers et des artisans au point de vue de l'hygiène et de la médecine légale." *Annales d'hygiène publique et de médicine légale*, 2d ser., XVII (1862), 104–190.

Virchow, Rudolf. "Die pathologischen Pigmente." *Archiv für pathologische Anatomie und Physiologie*, I (1847), 434, 461–466.

—— "Ueber das Lungenschwarz." *Archiv für pathologische Anatomie und Physiologie*, XXXV (1866), 186–190.

Vogt, Adolf. Die allgemeine Sterblichkeit und die Sterblichkeit an Lungenschwindsucht in den Berufsarten. *Zeitschrift für Schweizerische Statistik*, XXIII (1887), 249–297.

Wade, T. W. Report of an Inquiry . . . among Slate Quarrymen and Slate Workers in the Gwyrfai Rural District. Ministry of Health, Report 38. London, 1927.

Waters, C. M. An Economic History of England, 1066–1874. Oxford, 1925.

Watson, A. W. An Account of an Investigation of the Sickness and Mortality Experience of I.O.O.F. Manchester Unity. Manchester, 1903.

Weber, H. H. "Zur Methode der Analyse technischer Lösungsmittel." *Arbeiten aus dem Reichsgesundheitsamt*, LXIII (1931), 631–636; (with W. Koch) LXVI (1933), 79–82.

—— "Einfache Nachweise und Bestimmungsverfahren giftiger Gase, Dämpfe, Rauche und Staube in der Fabrikluft." *Zentralblatt für Gewerbehygiene*, XXIII, new ser., XIII (1936), 177–180.

Weigels, Christoff. Ständebuch von 1698 . . . mit beigedruckter Lehr und mässiger Vermahnung von P. Abraham a Santa Clara. New ed., ed. by W. Langewiesche-Brandt. Ebenhausen bei München, 1936.

Weinberg, W. "Die Sterblichkeit der Aerzte in Würtemberg."

Würtembergische Jahrbücher für Statistik und Landeskunde. 1897.

Wepfer, F. I. Observationes medico-practicae de affectibus capitis internis et externis. Schaffhausen, 1721.

Westergaard, Harald. Mortalität und Morbilität. 1st ed., Jena, 1882; 2d ed., Jena, 1901.

West Virginia. Annual Reports of the State Compensation Commissioner; Workmen's Compensation Fund. Charleston, West Virginia.

Wheeler, R. V., and D. W. Woodhead. The Lighting Power of Flame Safety Lamps. British Safety in Mines Research Board, Paper 40. London, 1927.

Winslow, C. E. A., and L. Greenburg. "A Study of the Dust Hazard in the Wet and Dry Grinding Shops of an Ax Factory." *Public Health Reports,* XXXV (1920), 2393-2401.

White, R. Prosser. Occupational Affections of the Skin. London, 1915; 4th ed. (under the title, The Dermatergoses or Occupational Affections of the Skin), London, 1934.

Whitney, Jessamine S. Death Rates by Occupation. National Tuberculosis Association. New York, 1934.

Wieck, E. A. The American Miners' Association. New York, 1940.

—— Preventing Fatal Explosions in Coal Mines. New York, 1942.

Yant, W. P., E. Levy, R. R. Sayers, and Others. Carbon Monoxide and Particulate Matter in the Air of Holland Tunnel and Metropolitan New York. Bureau of Mines, Report 3685. Washington, 1941.

Yant, W. P., H. H. Schrenk, R. R. Sayers, and Others. "Urine Sulphate Determination as a Measure of Benzol Exposure." *Journal of Industrial Hygiene and Toxicology,* XVIII (1936), 69-88.

Zangger, Heinrich. "Vergiftungen." In *Diagnostische und therapeutische Irrtümmer und deren Verhütung,* ed. by J. Schwalbe. Leipzig, 1924.

Zenker, F. A. "Ueber Staubinhalationskrankheiten der Lungen." *Deutsches Archiv für klinische Medizin,* II (1866), 116-172.

Index

Accident compensation, 48; nonprofit organizations for compensating workers, 45; miners first to receive, 221; for accidents en route to and from work, 271
Accident—compensation laws, 269
Accident Insurance Associations, German, 86; *see also* Workmen's Compensation
Accident prevention, 147-52, 268; protection against injury from tools in antiquity, 3; first act making provisions for, 24; institutions in factories for promotion of industrial hygiene, 94-111; safety committees and works committees, 94-99; safety directors and engineers, 99; education, propaganda, safety first, 150-52; books written by factory inspectors, 153; diminution in fatal accidents due to, 271; *see also* Factory inspection; Labor laws; Mine accidents; Mine inspection; Mine ventilation; Safety; Ventilation
Accident Prevention Week, 152
Accidents, protection against, in antiquity, 3; fatal, 49, 53, 111, 269, 270; responsibility for, 148; statistics, 264-76; effect of governmental control over circumstances leading to, 279, 282; *see also* Dangerous trades; Mine accidents; Workmen's compensation
Ackermann, J. C. G., 14, 168, 195, 280
Act for the Security of Collieries and Mines . . . 1800, 224
Act to Amend the Law Relating to Certain Factories and Workshops, 26
Act to Make Further Provisions for Regulation of Cotton Cloth Factories, 26

Actuarial Society of America, 187
Adelmann, Georg, 183
Adler-Herzmark, Jenny, 156, 158
Advisory Panel on Ophthalmology, 37
Agricola, Georg, 8, 100, 215, 232, 234, 280; quoted, 194
Air, measurement of harmful substances in, 128; detection and determination of injurious gases, 132-39; allowable limit of contamination, 138
Air Hygiene Foundation, 92, 206
Air line respirator, 144
Air space requirement per person in factory, 32
Aitken, J., 131
Alabama, General Laws contain no protection for workers, 68
Alberti, Michael, 12
Albrecht, H., 154
Alibert, J. J., 215
Alkali Works Regulations Acts, 28
Allegheny Ludlum Steel, 92
Allergy, 217
Allgemeine Arbeiter Kranken- und Unterstützungskasse (Vienna), 189
Allgemeiner Knappschaftsverein in Bochum, 190
Althorp Act of 1833, 34
Almaden, Spain, mercury mines, vii
Aluminum Company of America, 92
American Association of Industrial Physicians, 108
American Engineering Standards Committee, 92
American Federation of Labor, many leaders back free enterprise, 84
American Foundation of Occupational Health, 92
American Medical Association, 93, 108; opposition to Wagner-Murray-Dingell Bill, 80
American Miners' Association, 81

INDEX

American Museum of Safety, New York, 55
American National Safety Council, 141
American Society of Heating and Ventilating Engineers, 93, 182
American Standards Association, 92
Amman, Jost, 10
Ammonium persulfate, eczema caused by, 219
Andrae, A., 185
Andrewes, F. W., 200
Andrews, J. B., 54, 55, 229
Anilin poisoning, 277
Animal experiments to determine injurious gases, inconclusive, 137
Ankylostomiasis, 254-55
Anthracite Agreement of 1946, 82
Anthracite Coal Operators, 83
Anthracite Mines, higher dust concentration, 205
Anthracosis, 196, 197
Anthrax infections, 27, 28, 29; decrease in war years, 277
Appalachian Agreement of 1941, 82
Apprentices, legislation in England protecting, 22; for sea duty, 56
Arbeitsschutz, 1925, 151
Arbitration, 24
Archers, protection of thumb and hand, 3
Arianus, 208
Aristotle, 4, 208
Arizona, no legal control of industrial hygiene, xi; no rules for health of workers, 70; Industrial Commission, 61
Arkansas, urban population, 60
Arlidge, J. T., 154, 168, 200, 216
Arm-protecting plates, 3
Arnold, 198
Arsenic extraction, regulation of manufacture of, 27
Artisans, Ramazzini's treatise on the diseases of, 11
Asbestos, effect on lungs, 207
Ashley, Lord, 23, 224
Assiette au Beurre, 76
Association of British Chemical Manufacturers, 134

Association of Certifying Surgeons, 91
Association of Iron and Steel Electrical Engineers, 91
Association of Life Insurance Medical Directors, 187
Association of the Chemical Industry (German), 45
Association of Viennese Sickness Insurance Funds, 277
Associations and organizations, cooperating in industrial hygiene, 75-93; international, for labor protection, 87-90; national, 90-93
Atkinson, W. N., and J. B. Atkinson, 248
Aub, J. C., L. T. Fairhall, and others, 172
Austria, use of white lead in interior of buildings prohibited, 76; regulation of factories making phosphorus matches, 119; investigation of lead trades, 161; periodical on industrial hygiene, 176

Baader, E. W., and E. Holstein, 173
Bacon, Roger, 209
Baden, medical factory inspector, 106
Badham, Charles, 205
Bakehouse Act, 24
Baker, Robert, 117, 155
Bakers, skin diseases, 216, 218
Ballard, J. W., H. J. Oshry, and H. H. Schrenk, 164
Barclay, A. E., 205
Bassoe, J. J., 212
Bassoe, Peter, 55
Bateman, Thomas, 215
Bathing installations for miners, 222, 226, 227, 255
Bauer, Stefan, 88
Bavaria, control of child labor, 41; regulation for protection of workers, 42; medical factory inspector, 106; regulation of quick-silvering of mirrors, 119, 125, 127
Beck, K., and P. Stegmüller, 160
Behnke, A. R., 214

INDEX

Belgium, work by medical factor inspector Glibert, 157
Bell, J. H., 243
Benson, Thomas, 194
Benzol indicator, 137
Berger, Wilhelm, 217
Bergsucht, see Miners' phthisis
Bering, E., 217
—— and E. Zitzke, 219
Berlin, rule re manufacture of incandescent lamps, 119
Berlin Conference of 1890, 87
Bert, Paul, 211
Berufsgenossenschaften, 44, 45, 49
Betriebsrätegesetz, 97
Biram, B., 234
Birth certificates, 117
Bismarck, Otto von, ix; workmen's insurance legislation initiated by, 51
Bituminous Coal Mines Agreement, 98
Blackdamp in mines, 239, 246
Bladders, as protection against dust and fumes, 4, 139, 194
Blaschko, A., 216
Blänsdorf, E., 172
Blasting, in mines, 235, 236
Bloch, B., 217
Bloomfield, J. J., 162, 206
—— V. M. Trasko, and others, 98, 99, 108
Boas, Ernst, xiv
Boats, inspection, 56
Bohemia, German miners and skilled workers in, 221
Böhm, E., 116
Böhme, A., 204
Boiler inspectors, 71
Borellus, J. A., 144, 209
Boulin, M., 88
Bowditch, M., 65, 166
Boycott, A. E., 159, 212
Boyd, R. N., 223, 225
Braceland, F. J., 166
Breathing equipment for rescue work in mines, 253
Breton, J. L., 76
Brezina, E., 141, 156
Bridge, J. C., 101

Bridge building, caisson installation in, 210; illness of workers, 211
Bristons, Dr., 154
British Bakers' Union, 218
British Industrial Safety First Association, 151
British Institution of Civil Engineers, Committee on Regulations for . . . Work Carried Out in Compressed Air, 213; *Report*, 215
British National Health Act, 123
British Royal Society of Arts, 140
Britton, R. H., 162, 192
Brockmann, C. H., 132, 196
Bromley, J. F., 205
Bronce, F. A., 172
Brouardel, Paul, 76
Brown, J., 52
Brundage, D. K., 162, 192
Bruns, Hayo, 255
—— Löbker and, 160
Brunton, W., 234
Bryson, S., 204
Bubbe, Joannes, 195
Buck, Albert H., 53
Buddle, John, 86, 224, 234, 241, 252
Bundesrats Verordnungen, 43, 44, 147
Burke, W. J., 65, 165
Burton, C. C., 212
Butchers, death from tuberculosis, 184
Butler, T. H., 244
Büttner-Wobst, W., 207

Cains, John, 9
Caisson sickness, 55; best methods of avoiding, 212; publications concerning, 214
Caisson work, 208-15
Calder, J., 154
California, University of, 143
Cammack, E. E., 187
Cancer, mule-spinners', 36, 160, 167; of the lung among chromate workers, 77
Cap lamps for miners, 245
Carbon bisulphide, 166
Carbon monoxide, protection against, 137, 143; books on poison-

Carbon monoxide (*Continued*)
ing caused by, 173; indicators, 246, 247
Casper, J. L., 184
Certifying surgeons, 24, 101, 117
Chadwick, Edwin, 100
Chemical analyses, 136, 137
Chemical industry, foundation of, 16; sickness funds: work of factory physicians, 105
Chemical Works, British Regulations for, 133
Chevallier, A., 215, 216
Child labor, age certification, 24; apprentices for sea duty, 56; underground work prohibited, 56, 224; exclusion from dangerous work, 116-18, 120
—— Germany: regulation of, 41, 45
—— U.S., 57
Child labor legislation, 29, 281
—— England, 22 ff.
—— U.S.: state laws, 66 f.
Childs, S. B., 205
Chimney Sweep Act of 1788, 22
Christ, A., 212
Chromates, cancer caused by, 78
Cities, unhygienic conditions, 16
Civil Aeronautics, labor hours, 57
Cleanliness, regulations re, 24, 28, 126
Coal Act of 1911, (English), 240
Coal dust, methods of preventing explosions, 230, 248, 249; role in mine disasters, 247
Coal fumes, protection against, 7, 13
Coal mine explosions, *see* Mine accidents
Coal mine inspectors, *see* Mine inspectors
Coal miners, *see* Miners
Coal mines, poisonous gases, 9; compulsory insurance of workers, 49; production, 196, 228, 238, 260; dust, 198, 234-37; regulations of, in Ruhr District, 222; acts providing for inspection of, 225; wetting to decrease dust in air, 226, 230, 235, 236, 249; British government's right to consolidate, 227; nationalization of, in Great Britain, 228; rock-dusting in, 230; safety codes, 231, 263; only way to make American, as safe as those of Europe, 232; dangers and their control, 232-52 (*see also* Mine accidents); depth, 232, 260; mechanical installations, 233; blasting a source of dust, 235; methane firedamp, 238 ff., 246; indicators of gases, 246; explosives used, 247; legislation, 252; results of mine hygiene, 255-63; technical, economic, and legal differences between countries, 260; effect of mechanization on accidents, 261; ventilation, *see* Mine ventilation
Coal Mines Acts, British, 96, 248
Collis, E. L., 158, 159, 196, 201
—— and Major Greenwood, 17, 25, 186
Commerce, interstate, *see* Interstate commerce
Committee on Cornish Mines, recommendations re dust, 236
Committee on the Use of Lead in Potteries, 78
Compressed air, used in diving apparatus: in building shaft in coal mines, 209; in construction of tunnel, 210; method of avoiding ill effects of working in, 212; legislation protecting workers, 213
Compressed air illness, among divers and miners, 210
Congresses for occupational diseases, 177
Congress for Labor Legislation, Paris, 1900, 87
Congresso Nazionale per le Mallatie del Lavoro, 178
Consumption, *see* Phthisis; Respiratory diseases; Tuberculosis
Cook, Warren A., 74, 138
Cooke, W. E., 207
Cornish miners, exposure to stone dust, 235; protection, 236
Cornish mines, 201
Cotton factories, acts regulating, 26, 29; first in U.S., 51

INDEX

Crosson, J. W., 213
Cumberland, depth of collieries, 233
Curschmann, F., 105
Cutter's guild, 5

Dallavalle, J. M., 162, 206, 207
Dammer, O., 154
Dangerous substances, elimination of, 113-16
Dangerous trades, control of industrial damages, 18; power to establish special rules for, 25; protection of workers in, 32; law governing, 43; interest in, in U.S., 53, 54; exclusion of especially endangered groups from, 116-23; British rules concerning, 119; value of medical examinations, 120, 121; increasing attention paid to, 155; see also under Accidents; Factories; Mines; Occupational diseases
Davenport, Sara I., 163
Davis, G. G., 206
Davis, John, 245
Davy, Sir Humphry, 224
Davy safety lamps, 224, 240; candle power, 242
Death rate, see Mortality rate
Decompression illness, 210-15
Degea, A. G., 137
Degea "Kohlenoxyd Anzeiger," 247
Dejerine-Klumpke, Madame, 171
De la Beche, Henry, 253
Delaware, population, 60
Delmege, J. A., quoted, 5
Delpech, A., 166
Denmark prohibited production of yellow phosphorus matches, 113
Departmental Committee (British) on Compensation for Injuries to Workmen, 79
Dermatoses, 216
Desoille, H., 156
Deutsche Gesellschaft für Gewerbehygiene, 91
Devoto, Luigi, 177
Disasters, see Accidents
Diseases, occupational, see Occupational diseases

Dithizon method for detecting lead in air, 136
Divers, protective devices, 143, 144, 208; compressed air illness, 210; use of helium in deep-sea diving, 213
Dixon, S. M., 228
Doehring, C. F. W., 54
Draeger, Bernhard, gas detector, 137, 247; circulating apparatus, 144, 145
Drinker, P., 131
—— and T. Hatch, 207
Dublin, L. I., 186, 268
—— and Vane, R. J., 186
Duchenne, A. E., 140
Du Pont de Nemours, 92
Düsseldorf, regulations for protection, 42
Dust, law re installation of devices to keep, away from worker, 25; protection against, 32; technical measures to eliminate dangerous, 33, 127; sampling, 128-32; counting, 131, 132, 282; publications of U.S. Public Health Service on, 162; research and studies, 165; kinds, 198, 203; in mines, 234-37; methods of rendering dust of drilling innocuous, 237; see also Coal dust; Rock-dusting
Dust barriers in mines, 249, 250
Dust counter, Owens', 131
Dust diseases, ix, 194-208; development of dust lung into phthisis, 195; pathologic findings on dust workers, 197; chemical composition of dust a factor in lung affection, 199; skin diseases caused by dust, 215; study and control, 282
Dust masks, 4, 18, 139-42, 237
Dyes, regulation of manufacture of, 27

Earlywine, J. L., 206
Eczemas, occupational, 217 ff.
Ehrlich, W. E., 166
Electric lamps for miners, 243, 245
Electroplating installations, test in examining, 134

324 INDEX

Elkins, H. B., 65, 166
Ellenbog, Ulrich, 7
Elper, E., 116
Emergency Committee in Aid of Displaced Foreign Medical Scientists, xiv
Employers, liability for workers' accidents, 48 (*see also* Workmen's compensation); combat protective measures and liability laws, 147
Employers' associations, 86
Employers' Liability Act, U.S., 57
Employers' Liability Act of 1880, British, 26, 48, 49, 270
Enameling of iron plates, regulation, 27; health examination of workers, 120
Encyclopedists, new view of life initiated by, 15; influence, 280
Engel, H., 160
—— and V. Froboese, 160
England, control of industrial and mining accidents and diseases, viii, ix; capitalist system gave rise to working class, 7; industrial strides, 15; economic changes, 17; factory legislation, 22-37; safety committees and works committees, 94-96; factory physicians, 100-104; laws regulating employment of women and children, 24 ff., 116 ff.; rules concerning dangerous trades, 119; research on health of workers, 154; works by medical factory inspectors, 157; periodical on industrial hygiene, 177; works on toxicology, 181; mines and miners, 223 ff.; lead in labor protection and industrial hygiene, 26, 27, 280
—— Home Office: cooperation with employers' associations, 86
—— Ministry of Labour and National Service: *Factory Orders*, 30
—— Secretary of State: power to make special regulations for safety, health, and welfare in factories and workshops, 26-34 *passim*
Erdman, A. F., 13
Esch Law, 54, 56, 113
Eulenberg, Hermann, 168, 198, 216

Exhaust devices, 33, 53, 128, 282
Exhibitions, industrial hygiene promoted by, 178
Explosions, *see* Mine accidents
Explosions in Mines Committee, 249

Factory and Welfare Advisory Board (British), 35
Factory and Workshop Acts, England, *see* Factory legislation
Factories, ventilation, 31, 32, 33, 67, 282; investigations into conditions, 54; institutions in, for promotion of industrial hygiene, 94-111; movement to spread information on safety and industrial hygiene into, 150; *see also* Industries
Factory hygiene, gap between legislation and practice, xiv; hygienic-technical installations, 18; first aid installations, 29; importance of health insurance, 50 (*see also* Health insurance); institutions in factories for promotion of, 94-111; struggle of workers' organizations for improvement of, 75; movement to spread information on, 150; *see also* Industrial hygiene
Factory inspection, 153; *see also* Accidents, prevention
—— England, 33-37; consultation among specialists, 35; committees and preparatory conferences, 36
—— Germany, 45-48
—— U.S., 71-73; inferior to that in Europe, 108
Factory inspectors, England, appointment, 23; authorized to take samples in every factory, 30; duties, 33; first woman, 34; qualifications, 34; relations between management and, 35; specialized, 35
—— Europe: lifelong economic security, 71
—— Germany, 40, 43; civil service status, 45; qualifications: duties, 47
—— U.S.: civil service status, 71; not so well trained as European, 108; less carefully selected, 161

——— medical, x, 28, 34, 37, 47, 71; investigations: publications, 156 ff.; first, with years of their appointments, 156

Factory legislation, England, 22-37; industries other than textile first included, 25; gradual horizontal extension, 31; *1802*, 31; *1819*, 22; *1831* and *1833*, 23; *1844*, 23, 31, 117; *1864*, 25, 31, 32, 34; *1867*, 17, 25, 32; *1878*, 25, 118, 126, 147; *1883*, 26, 32, 126; *1891*, 26, 119; *1895*, 27; *1901*, 28, 29, 32, 118, 125, 147; *1907*, 28; *1911*, 28; *1916*, 29; *1937*, 29, 32, 36, 79, 102, 147

Factory Orders, *1940*, 102, *1944*, 30
——— Germany, 37-51
——— U.S., 56-60

Factory physicians, 100-111; under National Socialist Party, 106; part-time, 106; field of, 111

Fairhall, L. T., 136, 162
——— J. C. Aub, and others, 172

Fair Labor Standards Act of 1938, 56

Falacz, J. M., 70
Faraday, Michael, 225, 247
Farmer, E., 148, 159
Farr, W., 188, 195
Fay, A. H., 256, 257n, 261
"Federal Mine Safety Code," 83
"Federal Mine Safety Standards," 83
Fehnel, J. W., 206
Feicht, F. L., H. H. Schrenk, J. B. Littlefield, and, 164
Felt hat industry, use of mercury nitrate in, 115
Filter boxes, 142
Finlaison, A. G., 189
Finland production of yellow phosphorus matches prohibited, 113
Firedamp in mines, 238, 246, 247
Fire gilding, 19, 115
First-aid installations in factories, 29
Fischer, H., 136
Fischer, Richard, 155
——— T. Sommerfeld and, 88
Flax workers, lung diseases, 20
Fleuss-Davis Protoapparatus, 145

Flints, grinding by wet method, 194
Flury, F., 172
Forbes, J. J., 231; quoted, 232
Ford, C. G., 165
Foremen, safety instruction, 150
Fork grinders, phthisis among, 195
Foster, C. LeN., 129
Fourcroy, A. F. de, 13
"Fourneau d'appel," 18
France, legislation against nuisance of vapors and waste substances, 5; forbade use of white lead in building trades, 76; ceased to use yellow phosphorus matches, 113; books on occupational diseases, 175; periodicals on industrial hygiene, 177; works on toxicology, 181
Franconia, Middle: regulation of quicksilvering of mirrors, 119, 125, 127
Frank, H., 212
Frank, Johann P., 38, 215
Frankenberg, S., 171
Fraser, J. A., 159
Frazer, J. C. W., 143
Frederick the Great, 40
Free enterprise, fight for in England, 23
Freie Gewerkschaften, 77
French Revolution, influence, 17, 280
Freudenberg, F., B. Heymann and, 190
Frey, Emil, 87
Friedrich, W., 88
Friendly Societies, 185
Froboese, V., H. Engel and, 160
Fugger family, mining enterprises, 7
Fumes, first protection against harmful, 7 ff.
Fur, use of mercury nitrate in cutting, 115; use of mercurial carrot in preparation of, prohibited, 116

Gafafer, W. M., 192
Galvanoplastic method, fire gilding replaced by, 115
Ganister mines, 201
Gardner, Leroy U., 206
Garment industry of New York, conditions investigated, 55

Gas bubbles in blood vessels of workers in compressed air, 212
Gas detector, 137
Gases, developing in mines, 9, 14; first protection against noxious, 18; technical measures to eliminate, 127; detection and determination of injurious, in air, 132-39; recognized by changes in flame of miners' lamps, 240, 246; indicators, 246; *see also* Methane
Gas masks, 142-43
Geissler, A., 185
Georgia, Labor Laws, 68
Gerbis, H., 156, 158
German Health Insurance Act, 105
German Association for Industrial Hygiene, 44, 77
Germany, doctrine that minerals in earth were property of state, viii; start of industrialism and capitalism, 6; economic changes, 17; factory legislation, 37-51; industrial development, 39; right of government to interfere in all matters, 39, 40; *Gewerbeordnung*, 42, 43, 44, 46, 126; *Arbeiterschutzgesetz*, 42; factory inspection, 45-48; police supervision of execution of labor laws, 46; war against dangers of lead paints, 76; Labor Protection Law of 1891, 96; works committees, 96-98; labor-management relations under National Socialism, 97; factory physicians, 104-8; prohibited manufacture of white phosphorus matches, 113; regulations of employment of women and young workers, 118, 120; medical examinations, 121; mine rescue apparatus, 144; Accident Prevention Week, 152; research and studies, 155; works by medical factory inspectors, 157; *Reichsgesundheitsamt, Arbeiten*, 160; investigations on public health and industrial hygiene, 160; books on occupational diseases, 174; periodicals on industrial hygiene, 176; works on toxicology, 180; accident-compensation laws, 48, 49, 269; fatal accidents by industries, *tabs.*, 272, 275; compensated cases of occupational poisoning, *tab.*, 277
—— *Bundesrat*, later *Reichsarbeitsminister:* right to issue protective labor rules, 43
Gewerbeordnung of 1869, Germany, 42, 43, 44, 46
Gewerbeunfallversicherungsgesetz, 48
Gewerbliche Nachtarbeit der Frauen, 88
Gewerbliche Berufsgenossenschaften, 44, 45, 49, 271
Gibbs, mine rescue apparatus, 145
Gilders, industrial hygiene for, 13, 19
Glass masks, 4
Glibert, D., vii, 156, 157, 173
Goadby, K. W., 171
—— T. M. Legge and, 171
Goethe, 37
"Goldene Bulle," right to mines in, ceded by German emperor, 220
Gold mines, dust content of air of South African, 237
Goldsmiths cautioned against poisonous fumes, 7
Goldwater, L. J., L. Greenburg, and others, 4n, 164
Gothaer Lebensversicherungsbank, 185
Götzinger, W. L., 195
Government control, effect upon accidents and occupational diseases, 279, 282; right of officials to issue regulations for hazardous industries, 281
Government ownership of minerals in earth, 220
Great Britain, prohibited use of white phosphorus matches, 113; mine rescue apparatus, 144; books on occupational diseases, 174; colliery disasters caused by explosions, *tab.*, 251; rescue work in mines, 252; annual death rates of mine workers, *tab.*, 257; by fatal

accidents, *tabs.*, 258; fatal accident in industries, *tab.*, 270; cases of occupational diseases, *tab.*, 276; compensated cases of lead poisoning, *tab.*, 277; *see also* England
—— Health of Munitions Workers Committee, 159
—— Home Office, 151
—— Industrial Fatigue Research Board, 148, 159
—— Industrial Health Research Board, 159
—— Medical Research Committee, 159, 160
—— Medical Research Council, 159, 160, 228
—— Mines Department, 226, 228
—— Ministry of Munitions: instructions for factory physicians, 160
—— Safety in Mines Research Board, 228, 243
Greenburg, Leonard, 63, 65, 131, 162, 206, 207
—— M. R. Mayers, A. R. Smith, and others, 164
Greenhow, Dr., 154, 197, 200
Greenwood, Major, 23, 159, 264
—— E. L. Collis and, 17, 25
Grinding trade, regulations, 32, 33, 42
Grotjahn, A., 200
Grout, J. L. A., 205
Grove, G. W., 231
Guilds, 4; no provision for protection of workers, 5; handicapped development of industry, 37; government regulation, 37, 39; old organization by restricted, 40
Guy, W. A., 186

Habeas Corpus Act of 1701, Scotch, 224
Haftpflichtgesetz of 1871, 48
Hahn, Martin, 130
Haldane, J. S., 129, 141, 155, 201, 212, 213, 228, 235; quoted, 244
Halfort, A. C. L., 20, 115, 215
Hallé, Noel, 254
Hamilton, Alice, vii-xi, 54, 55, 115, 116, 164, 166, 170

Hannover, A., 216
Hanson, W. C., 55
—— G. M. Kober, 170
Harrington, D., 163, 230
Hartmann, J., 260, 262
Harvey, V. K., 122
Haskon, L., 210
Hat industry, search for substitute for mercury nitrate used in, 115
Hatch, Theodore, 65, 164
—— P. Drinker and, 131, 207
Haupt, Gustav, 77
Hayhurst, Emery, 55
Health, task of state to supervise, in Europe, 84; industrial health service, 103, 104; methods for control of hazards, 112-46; task remaining to improve industrial, 282; *see also* Hygiene; Occupational diseases
Health and Morals of Apprentices Act of 1802, 22, 126
Health centers, periodic examination of workers by, 123
Healthful living, need for guidance in, 107
Health insurance, 50-51; early form, 100; Great Britain, 51, 80; Germany, 51, 80, 104; U.S., 108, 111, 193; compulsory, in sick fund, 104, 192, 277; demanded comprehensive statistics, 189, 193
Health of Munitions Workers Committee, British, 101, 148
Health rating of trades, 192
Hebburn colliery, first mechanical ventilation, 234
Hebenstreit, E. B. G., 38
Hebestreit, H., 106
Hedley, O. F., 109
Heiman, H., 65
—— L. Greenburg, and others, 164
Heinzerling, C., 169
Helium used in deep-sea diving, 213
Heller, R., 210, 212
—— W. Mager, and H. von Schrötter, 214
Helmont, F. M. van, 143
Helvetic Republic, 87
Henckel, J. F., 12, 194

328 INDEX

Henry IV, king of England, 223
Herport, A., 210
Hesse, 129, 141, 197, 234
Heymann, B., and F. Freudenberg, 190
Hides and skins, regulation for handling to avoid anthrax infections, 28, 29
Higgins, Edwin, 59, 128, 205
Highland Society, 189
Hildesheim, regulation for protection of workers, 42
Hill, L. E., 130, 159, 182, 214
Hill-Burton Bill, 80
Hippocrates, 4
Hirt, Ludwig, 168, 198, 216; quoted, 167 f.
—— and G. Merkel, 169
Hobbs, A. A., 207
Hoffman, Frederick L., 54
Hoffmann, Friedrich, 12, 13
Hohenheim, Theophrastus von (Paracelsus), 9, 10, 100, 194, 215
Holland, G. E., 21
Holstein, E., 107
—— E. W. Baader and, 173
Holtzmann, F., 156
Homework Law of 1914, Germany, 43
Hookworm disease, 254
Hoolamite detector for CO, 246
Hopcalite, 143
Horn, Generalleutenaut von, 41
Hose masks, 143
Houghten, F. C., 182
Hours of labor, see Working hours
Hudson Tunnel, 211
Hue, O., 232
Humboldt, A. von, 144
Humboldt, Wilhelm von, 40
Humidity, 31
Hungary, German miners and skilled workmen in, 221
Hyde, R., 100, 101
Hydrocyanic acid, test used in control of, 134
Hygiene, in mines, 7-11, 220-63; progress of legislation on general factory, 31 (see also Factory hygiene; Factory legislation); and physiology, 182; see also Industrial hygiene
Hygienic working day, see Working hours, shorter

Illinois, acts providing compensation for occupational diseases, 73; miners' agreement re mining legislation, 81; coal output, 228
—— Industrial Commission: rules, 70
—— Labor Department: Industrial Hygiene Section, 62
—— Mining Investigation Commission, 81
—— Occupational Diseases Committee, 55
Illnesses, scheme for statistics, 190
Imbert, 148, 149
Imbert-Gourbeyre, 215
Impinger, 131
Indiana, coal output, 229
Indians, work in mines, 14
Industrial accidents, see Accidents
Industrial danger, coordinated efforts against, 160
Industrial diseases, see Occupational diseases
Industrial dust, see Dust
Industrial Dust Hazard Panel, 35
Industrial Health Advisory Committee, 35
Industrial health service, aims, 103; financing of, 104; see also Health
Industrial hygiene, development, 3-21; ancient, 3-4; medieval, 4-7; 15th to 17th centuries, 7-11; 18th century, 11-15; 19th century scientific and economic evolution, ix, 15-21; importance of health insurance, 50 (see also Health insurance); organizations and associations cooperating in, 75-93; workers' unions, 75-86; interest of trade unions in, 79 ff., 85; miners active in field of, 81; international association, 87-90; national associations, 90-93; institutions in factories for promotion of, 94-111; safety and works committees, 94-

99; safety directors and engineers, 99; factory physicians, 100-111; World War II brought great advances in appointment of factory physicians, 110; research and studies: governmental, 153-66; nongovernmental, 166-79; progress in other sciences and its influence on, 180-93; toxicology, 180-81; physiology, 182; statistics, 182-93; special problems, 194-219; effect, 264-79; laws and regulations indispensable to practice of, 283; *see also* Factory hygiene; Mine hygiene; Occupational diseases
—— U.S., ix, 51-74; federal legislation and activities, 56-60; research, 58; state legislation and activities, 60-66; state laws, codes, and regulations, 66-71; factory inspection, 71-73; workmen's compensation, 73-74
—— legislation: development, 17; Great Britain, 22-37; protecting workers in compressed air, 213; Germany, 37-51; U.S., 51-71; on mines, 220-32, 252; *see also* Child labor legislation; Factory legislation
Industrial Hygiene Foundation, 92
Industrial Injuries Act of 1946 (British), 49
Industrial medicine, movement under way for, 108
Industrial nurses, 108, 109
Industrial physicians, *see* Factory physicians
Industrial Welfare Society, 101
Industries, in medieval England, 5; special regulations of dangerous, 27 (*see also* Dangerous trades); development on Continent, 37; working hours in, 125 (*see also* Working hours); fatal accidents by, in Germany, tabs., 272, 275; *see also* coal mines; Factories; Mines
Injuries, *see* Accidents
Ins, A. von, 198
Inspectors, *see* Factory inspectors; Medical inspectors; Mine inspectors
Institute of Industrial Hygiene (Germany), 91
"Instruction for Medical Examination of Lead Workers," 161
Insurance, compulsory in coal mining, 49; unemployment, in Great Britain, 51; *see also* Accident compensation; Health insurance; Sickness funds; Workmen's compensation
Insurance companies, safety inspectors study accident prevention and control of industrial diseases, 74; safety education programs, 151
Interferometer, 135
International Association for Labor Legislation, 87-89, 125, 172
International Association for Labor Protection, American section, 55
International Congresses for Rescue Work and First Aid, 178
International Congress of Hygiene and Demography, 55
Internationale Hygiene Ausstellung, 178
Internationale Uebersichten über Gewerbekrankheiten . . . , 156
International Exhibition for Health and Rescue Work, 178
International Congresses for Occupational Diseases, 177
International Congresses on Occupational Accidents and Diseases, vii, 177
International Labor Association, 87
International Labor Conference, 1925, 50; 1928, 95; 1935, 50
International Labor Office, xi, 29, 87, 89-90; research work and publication on industrial hygiene, 90; *Health and Occupations*, 173
International Ladies Garment Workers Union, health center for members, 80
International Medical Congress, 1880, 242

International Permanent Commission for the Study of Occupational Diseases, 177
International Silicosis Conferences, 206
Interstate commerce and transportation, federal acts re, 56
Irvine, L. G., A. H. Watt, and others, 202
Italy, books on occupational diseases, 175; periodicals on industrial hygiene, 176

Jacobs, M. B., 136
Jadassohn, J., 217, 219
Jaeger, F., 212
Jaeger, H., 217
Jaquet, A., 167
John, king of England, 223
Johns Hopkins University, 143
Johnson, I. Pratt, 202
Joint Industrial Councils, committees on safety, health, and welfare, 36
Joint Industry Safety Committee, suggested, 84
Joint Occupation Study 1928, 187
Jones, D. C., 234
Jones, R. R., 208, 213
Journals of industrial hygiene and occupational diseases, 175 ff.
Justice of the Peace, duties, 34

Kaiserliches Gesundheitsamt, 155
Kaiserliches Statistisches Amt, 190
Kalstrom, S. C., 212
Kammerer, J. F., 113
Kansas, no legal control of industrial hygiene, xi; no rules for health of workers, 70
—— Labor Department, 61
Karl IV, German emperor, 220
Katathermometer, 182
Katz, H., and L. Teleky, 171
Katz, S. H., 141
Kaup, I., 77, 129, 161, 200
Kendall, L. B., 170
Kentucky coal output, 229
Kircher, Athanasius, 4, 9
Kjaer, S., 273

Knights of St. Crispin, Order of the, 60
Kober, G. M., 54
—— and W. C. Hanson, 170
Koch, Robert, ix, 199
Koelsch, F., 76, 134, 156, 200, 204, 212
Koerting injector, 144
"Kohlenoxyd Anzeiger," 247
Konimeter, 130, 132
Koniscope, 131
Kossoris, M. D., 271, 273
Kotzé, Robert, 130, 202
Kraft, M., 154
Kranenburg, W. E. R., 156
Kreislaufgerät, 144
Krüger, E., 207
Kulman, Josef, 218
Kussmaul, A., 172

Laborers, *see* Workers
Labor law inspectors, qualifications, 72; *see also* Factory inspectors
Labor laws, federal, 56; state, 66-71; international cooperation urgent, 87; *see also* Accident prevention; Child labor legislation; Factory legislation; Industrial hygiene legislation
Labor organizations, *see* Trade unions
Labor protection in U.S., federal authorities in field of, 59
Labor-management relations, works committees created to deal with, 94, 96
Labor turnover, 124
Ladders, accidents in mines using, 233
Ladies Garment Workers Union, medical and dental institution, 81
Lamps, miners', 239; *see also* Safety lamps
Lane, R. E., 102
Laney, F. B., 128
Lanza, A. J., 59, 74, 128, 205
Laundries, 28
Laurion silver mines, 232
Lavoisier, Antoine L., 16
Layet, A., 174, 216

INDEX

Lead, in air, 129
Lead colic, 4
Lead industry investigation, 54
Lead paints, use in buildings forbidden, 76, 278; paints to replace, 114; poisoning from, 129
Lead poisoning, in potteries, 6n; protection against, 29; treatises on, 54; among painters, 129; literature on, 170 ff.; compensated cases, Great Britain, *tab.*, 277; workers' illnesses with disability, *tab.*, 278
League of Nations, *see* International Labor Office
Leclaire, Jean, 114
Legge, Thomas M., 28, 34, 79, 101, 122, 156, 158, 159, 200, 201, 277
—— and K. W. Goadby, 171
Legislation, *see* Labor laws and legislation
Lehmann, K. B., 129, 135, 180
Leipziger Ortskrankenkasse, 190
Levine, L., 166
Levy, E., 212, 214
Lewey, F. H., F. J. Braceland, and W. E. Ehrlich, 166
Lewin, L., 13, 173
Lewis, John L., 83
Liebig, Justus von, 16, 114
Life, duration of, in several professions and trades, 184 ff.
Lincoln, D. F., 52
Linden, Mrs. B. J. W., 172
Linné, Karl von, 195
Lippe, regulation to control dangers in pits for marl, loam, and clay, 41
Lippmann, O., 149
Lithographers, age at death, 184
Lithopone, 114
Littlefield, J. B., F. L. Feicht, and H. H. Schrenk, 164
Lochtkemper, J., 204
Löhneiss, E. G., 194
Lombard, H. C., 184
London, safety conference, 151
Longshoremen and Harbor Workers' Compensation Act, 74
Loriga, G., 156
Le Grand, Luc, 87
Lungs, cancer of, among chromate workers, 77; essential factor in development of chronic diseases, 202; effect of dust on, 194-208
Luongo, E. P., 122
Lussigne, 116
Lyell, Professor, 225, 247

McConnell, 93
McCrae, J., 202
McCready, Benjamin W., 52
Machinery, legislation re guards for dangerous, 24, 26, 67; regulations for construction of, 30; attempt to make safe, 147
Macklin, E. L., 204
Maclaren, Dr., 200
McMahon, M., 65
Mager, W., R. Heller, and H. von Schrötter, 210, 212, 214
Mai, Franz Anton, 38
Malard, 115
Malicious Injuries Act, 1769, 224
Management, relations between inspectors and, 35
Manchester Royal Infirmary, 167
Manchester Unity, 189
Man-engine, accidents in mines using, 233
Marsaut lamp, 240; candle power, 242, 243
Martial, 4
Masks, bladders, 4, 139, 194; glass, 4; need for industrial, 139; dust, 141, 142, 237; gas, 142-43
Massachusetts, child labor, 53; population per square mile, 60; child labor law, 71
—— Board of Health: report on the conditions affecting health and safety in factories, 54 f.
—— Bureau of Labor Statistics, 60
—— Labor Department, 61, 62, 65, 165
Match industry, use of yellow (white) phosphorus prohibited, 28, 56, 89, 113; law concerning manufacture of phosphorus, 43; tax on white phosphorus, 54, 113; use of white phosphorus investigated, 88; invention of "Swedish" matches;

332 INDEX

Match industry (*Continued*)
  use of phosphorsesquisulfide, 113;
  regulation of factories making
  phosphorus matches, 119; phosphorus found in air of factory,
  133; *see also* Esch Law
Mavrogordato, A., 203
Maximus, Valerius, 13
Mayer, R. L., 217, 219
Mayers, M. R., 65
—— L. Greenburg, and others, 164
Mecco, mine rescue apparatus, 145
Medical Adviser, 79
Medical examination, pre-examination for factory work, 117, 120; of workers not exposed to specific dangers, 123; periodic, in dangerous trades, 122; governmental supervision: value of pre-examination, 121; *see also* Factory inspectors, medical; *see also* Factory physicians
Medical inspectors of factories, 28, 34, 37, 47, 71; investigations: publications, 156 ff.
Medical police, Germany, 38
Mérat, F. V., 170
Mercurial carrot, resolution prohibiting use in preparation of hatter's fur, 116
Mercury, mining of, vii; replaced by silver nitrate in production of mirrors, 114
Mercury air pumps, 119
Mercury nitrate, search for method to replace, in hat industry, 115
Mercury poisoning, vii; from fire gilding, 1, 13, 14, 115; literature, 172
Merewether, E. R. A., 101, 102, 207
Méricourt, Le Roy de, 210
Merkel, G., L. Hirt, and, 169
Metal cage, accidents in mines using, 233
Metal grinders, pneumoconiosis of, 204
Metallic dust, 198; *see also* Dust
Metalliferous miners, sixteenth century, viii; mortality from respiratory diseases, 195; fatal accidents in Great Britain and Prussia, *tabs.*, 258
Metalliferous Mines Act, 225
Metal workers, Hippocrates, reference to, 4; cautioned against poisonous fumes, 7
Methane, explosions, 9; danger of, in mines, 238 ff.; explosive concentration in air, 239, 241; recognized by changes in flame of lamps, 240; safety maximum in return air, 241
Metropolitan Life Insurance Company, 151, 186; death rate by age groups, *tab.*, 266
Mica dust, effect on lungs, 207
Michigan, University of, classes in health and safety, 85
Middle Ages, industrial hygiene, 4-7
Middleton, E. L., 35, 204
Miller, Genevieve, 52
Mills, injury to health and life, 19
Milroy lectures, 196, 204
Mine accidents, in Prussia, 222, 233
Medical Research Council, Special Reports, 205
"Medical Series," 92
Medical Society of the State of New York, 51
Medical supervision in industry, 30, 100-111
Medical treatment, fees for, 49
Meinel, F., 197
Mellon Institute, Pittsburgh, 92
Mine accidents, explosions, 224 ff., 238-41; investigation: number of fatalities, 225; control of explosions, 230, 239, 247-52; role of coal dust, 247; experimental galleries, 248, 249; falling rocks or coal, 250, 262; rescue work, 252-54; decreasing rate of fatal, in coal and metalliferous mines, 256; effect of mechanization on, 261; most efficient factors in diminishing, 262; fatal, in U.S., x, *tab.*, 259; annual death rates, *tabs.*, 257-59; fatality rates by states in U.S., 259; mortality rate in U.S. higher than in Europe, 262

INDEX 333

Mine Disaster Commission, Prussian, 247, 248
Mine inspection, acts providing for, of coal mines, 96, 225
Mine inspectors, 224, 231; U.S., 83, 229, 231; qualifications higher in Europe than in U.S., 262
Mine physicians, German, 100
Mineral dusts, 198; all not equally dangerous, 203; *see also* Dusts
Mineral resources, state ownership, viii, 220, 223; ownership returned to landowner, 223
Mine Rescue Apparatus Research Committee, 145
Miners, 220-63; compulsory insurance, 49; number killed annually by accidents, 53; active in field of mine hygiene, 81; demand minimum health and safety standard, 81; changes in lungs of coal and ore miners, 196; pneumoconiosis of, coal miners in Ruhr District, 204; pulmonary disease, 205 (*see also* Miners' phthisis); first to receive accident compensation, 221; bathing accommodations, 222, 226, 227, 255; strikes and struggles, 222; social condition, 223; provision for punishment of, 224; society to promote hygiene of, 224 (*see also* Hygiene); facility for drying clothes, 226; welfare work, 227; mortality from respiratory diseases among Cornish: exposure to stone dust, 235; protection, 236; silicosis among South African, 237; technical education, 252; decrease in mortality of coal miners, 256; annual death rates in Great Britain, *tab.*, 257; annual death rates by accidents in Great Britain and Prussia, *tabs.*, 258; in U.S. *tab.*, 259
Miners' diseases, *see* Ankylostomiasis; Miner's phthisis; Miners' nystagmus; Silicosis
Miners' National Association, 225
Miners' nystagmus, 243 ff.
Miners' Nystagmus Committee, 242, 244

Miner's phthisis, 194 ff.; danger increased by power-driven tools, 234-37
Miners' Welfare Fund, 227
Mines, hygiene, 7-11, 220-63; Act of 1890–1891 dealing with, U.S., 56; detection of toxic or suffocating gases, 132; rescue apparatus, 144; legislation, 56, 117, 220-32, 252; *Gewerken:* societies, 221; technical regulations, 222; standard of safety, 225, 226; use of water jets or sprays for machine drilling, 226, 235, 236, 249; supervision in U.S., 228; safety stations, 229; dangers and their control, 232-52; depth, 232; dust, 234-37 (*see also* Coal dust; Dust); blasting a source of dust, 235; dust in air of gold mines, 237; study of methods to render-dust of drilling innocuous, 237; ore mines, 238; illumination, 242 ff. (*see also* Safety lamps); dust barriers, 249, 250; rescue work, 252-54; results of mine hygiene, 255-63; governmental control greater in Europe than in U.S., 262; *see also* Coal mines
Mine Safety Appliances Company, 131, 137
Mine safety committees, 82
Mines and Collieries Act of 1842 (British), 117
Mines (Coal) Regulation Act (British), 225
Mine ventilation, 8, 9, 15, 53, 225, 234, 240, 262; the best way to avoid explosions, 239
Mining industry in Germany and in England, viii, 220-28
Ministers, age at death, 184
Minot, A. S., 172
Miquel, 128
Mirror makers, 19
Mirrors, rules for factories using mercury, 42; mercury replaced by silver nitrate in production of, 114
Mississippi, no workmen's compensation law, 73

334 INDEX

Moir, C. W., quoted, 211
Moir, J., 202
Montana, law re working hours: wages, 68; workmen's compensation, 73
Morbidity of occupations, 183 ff., 189
Morley, John, 167
Morrah, Dermat, 226; quoted, 227
Morris, G. E., 166
Mortality rate, in U.S. in 1945, 111; statistics, 182-93; in occupations, 183 ff., *tab.*, 267; decrease in, of English male population, 256, 257, 264; according to social class, *tab.*, 265; percentage of deaths by age groups, *tab.*, 266; fatal accidents in industries, Great Britain, *tab.*, 270; fatal accidents by industries, Germany, *tabs.*, 272, 275; U.S., *tabs.*, 274, 275
Morveau, Guyton de, 114
Moskowitz, S., 165
Motor carriers, hours regulated, 56
M.S.A. combustible gas indicator, 246
M.S.A. hydrogen sulphide detector, 246 f.
Mueseler lamp, 240
Mule-spinners' cancer, 36, 160, 167
Müller, Richard, 88
Murder, use of vapors and fumes for, 13
Murray, Montague, 207
Museums, industrial hygiene promoted by, 178
Musschenbroeck, van, 211

National Association of Manufacturers, 108
National Bituminous Wage Agreement of 1946, 82
National Committee for Conservation of Manpower, 57
National Conference on Occupational Diseases (U.S.), 55
National Conference on Labor Legislation (U.S.), Dec., 1945, Committee on Safety and Health and Workmen's Compensation, 57

National Health Insurance Medical Research Committee (British), 160
National Insurance (Industrial Injuries Act), 1946 (British), 49
National Insurance Act, 1911 (British), 51
National Health Service Act (British), 193
National Tuberculosis Association (U.S.), 188
National Silicosis Conference of 1936 (U.S.), 206
National Safety Council (U.S.), xi, 55, 91, 111, 150
National Safety Congress (U.S.), 91
National Union of Quarrymen (British), 78
National Union, miners', organized (British), 225
Nef, J. U., 223, 238
Neison, F. G. P., 185, 189
Neisser, E. J., 156
Nelson, W. T., 205
Netherlands, study of occupational diseases, 175
Neufville, W. C. de, 184
Nevada, population per square mile, 60; labor laws, 68, 147
Newman, George, 159
New York City, investigation of sweatshops, 80
New York State, population, 60
—— Factory Investigating Commission, 55
—— Industrial Hygiene Laboratory, 136
—— Labor Department, 60, 61; industrial health program, 64; Board of Standards and Appeals, 64; Bureau of Factory Inspection, 62; Bureau of Labor Statistics, 62; Division of Industrial Hygiene and Safety Standards, 62, 63; publication, 63, 164; Division of Industrial Safety Service, 64, 65; Factory Investigating Commission, 62; Industrial Board, 63; Workmen's Compensation Board, 64
—— Labor Law, 67

INDEX 335

Nicloux, M., 173
Nieden, A., 242, 243, 244
Nitrogen bubbles, in blood of workers in compressed air, 211; in tissues, 212
Norris, E. H., 111
North Carolina labor laws, 69
North Dakota, urban population, 60
North German Confederation, *Gewerbeordnung*, 42, 43, 44, 46
Notice of Accidents Act (British), 1906, 28
Nurses, industrial, 108, 109
Nystagmus, miners', 243 ff.

Occupational accidents, *see* Accidents; Mortality rate
Occupational diseases, first measure taken to avoid, 3; first articles published in England, 10; control of industrial damages, 18; lung diseases of flax workers, 20; attending physician required to report, 30; compensation, 45, 49, 73; diagnosis of case necessary for compensation, 50; interest in, in U.S., 53; states' provisions for, 73; federal laws, 74; instruction in prevention, 150; books on, 173-75; Great Britain, *tabs.*, 276, 277; effect of governmental regulations, 279, 282; clinical studies, 282; *see also* Ankylostomiasis; Anthrax; Cancer; Lead poisoning; Nystagmus; Phthisis; Poisoning; Silicosis; Skin diseases
Occupations, mortality of, 184 ff., *tab.*, 267; *see also* Mortality rate
—— dangerous, *see* Dangerous trades
Oesterlen, F., 185
Oettingen, W. F. von, 162
Oklahoma, no legal control of industrial hygiene, ix
Oldendorff, A., 185
Oliver, Sir Thomas, 54, 153, 168, 216
O'Neill, Charles, vii, 54
Oppenheim, M., K. Ullmann, J. H. Rille and, 219
Ore miners, changes in lungs of, 196

Ore mines, 238
Organic chemistry, 16
Organic dust, 198; *see also* Dust
Oshry, H. J., H. H. Schrenk, J. W. Ballard, and, 164
Overcrowding, 31, 32
Owen, Robert, 23
Owen-Illinois Glass, 92
Owens, J. S., 131, 132
Oxygen, self-containing, apparatuses, 144-146, 253; use of, during decompression, 213

Painters, lead poisonings among, 29, 129, *tab.*, 278; *see also* Lead poisoning
Paints to replace white lead, 114
Palmer, G. T., apparatus of, 131, 132
Pansa, Martin, 9, 100
Papin, Denis, 209
Paracelsus, *see* Hohenheim, Theophrastus von
Paris Society for the Encouragement of Industry, 115
Parson, A. C., 218
Patissier, P., 18, 19, 140, 168, 280
Patton, J. R., 207
Pearce, S. J., H. H. Schrenk and, 164
Pearson, George, 196
Pediculosis, 117, 218
Peel, Sir Robert, 22; quoted, 269
Pennsylvania, child labor, 53; population, 60; coal output, 228; mine inspectors, 229
—— Anthracite Mine Inspectors, 53
—— Department of Health, 206
—— Labor and Industry Department, 166
Peripneumonia, 194, 195
Permanent Commission for the Study of Occupational Diseases, 177n
Permanent International Committee for the Study of the Medicine of Accidents, 177, 178
Personal hygiene, importance of morals and, 264
Pertussi-Gastaldi test, 134
Pettenkofer, M. von, 169

Pfannmüller, caisson constructed by, 210
Phemister, D. B., 212
Philipps, William, 209
Phosphornecrosis, 113; books on, 173
Phosphorsesquisulfide, matches utilizing, 113
Phosphorus, in air of match factory, 133; *see also* Match industry
Phthisis, installations to control grinder's, 20; miner's phthisis, 194 ff., 234 ff.; development of dust lung into, 195; main causes, 196 ff.
Physicians, factory physicians, 100-111; mine physicians, 100; mouthpieces worn during plague epidemics, 139; important role of private, 167; age at death, 184; *see also* Factory inspectors, medical; Medical examination; Medical inspectors; Surgeons
Physiology, hygiene and, 182
Pierracini, G., 148
Pittsburgh Experimental Station, 250
Plato, 4
Playfair, Lyon, 253
Pliny, 4, 208
Pneumatogen apparatus, 146
Pneumatophor, 144
Pneumoconiosis, *see* Dust diseases; Miner's phthisis
Poisoning, legislation requiring reporting of cases caused by lead, phosphorus, or arsenic, 27; worst symptoms of chronic, becoming rarer, 51; cases increased by war production, 277; anilin, 277; compensated cases of occupational, in Germany, *tab.*, 277; occupational workers' illnesses with disability, caused by lead poisoning, *tab.*, 278; *see also* Carbon Monoxide; Lead poisoning; Match industry; Mercury poisoning
Poisonous substances, three stages in intake of, 122
Poisons, exclusion of women from work with particular, 118; re toxic limits of, 138; increased number of industrial, 155; U.S. Public Health Service Bulletins on, 162; research and studies, 165
Pol, B., and T. J. Watelle, 210, 211
Poland, German miners and skilled workers in, 221
Police, Factories, etc., Act of 1916, England, 29
Police physician, 38
Political action, union attitude toward, 86
Porro, F. W., 207
Posture, 195
Pott, Percival, 22
Potteries, examination of women and young workers, 120
Potters, use of wet glazes recommended, 15
Potton, F. A. F., 215
Pratt, E. E., 62
Preussisches Grubensicherheitsamt, contest for dust protection, 141
Prevention of accidents, *see* Accident prevention
"Preventive Engineering Series," 92
Price, G. M., 55, 62, 170
Priest Bill, 80
Prinzing, F., 186
Prodan, 136
Professions, mortality of, 184 ff.
Prophylaxsis, basis for, 11; periodic examination controls efficiency of, 123
Protective devices, 139-42
Proxylen apparatus, 146
Prussia, effort to promote industries, 40; poor condition of children in Rhineland, 41; measures to protect workers, 42; factory inspection, 45; qualifications, 47; medical inspectors of factories, 47, 106; mine law of 1905, 96; search for a mine rescue apparatus, 145; reports of medical inspectors, 156; scheme for illness statistics in the Rhineland, 190; coal mines 196; supervision of mines, 223; colliery disasters caused by explo-

INDEX   337

sions, *tab.*, 251; rescue work in mines, 253; fatal mine accidents, 258; rate of coal mine accidents compared with those of U.S., 260
—— Minister of Commerce: competition for methods of rendering dust of drilling innocuous, 237
Public health, responsibility for, in England and Germany, 80; examination of workers not exposed to specific dangers, 123
Pulmonary disease, in miners, 205; *see also* Phthisis; Respiratory diseases; Silicosis; Tuberculosis
Purdy, Dr., 201
Pütsch, A., 154

Qualifications for General Labor Law Inspectors, excerpt, 72
Quartz, interaction of, with coal, shales, and other strata, 205
Quartz dust, most dangerous dust, 132, 197; mortality from exposure to, 201; only dust that generates a specific disease, 203
Quicksilver, *see* Mercury

Railroads, safety appliances and inspections, 56
Ramazzini, Bernardo, 4, 14, 168, 194, 215, 280; and his successors, 11-15
Raymond, R. W., 53
Red lead factories, health examination of workers, 120
Rehn, L., 167
Reichel, F., 153
Reichsarbeitsblatt, 151, 190
Reichsarbeitsverwaltung, 152
Reichsgesundheitsamt, 155
Renatus, F. V., 208
Renaut, J., 171
Research and studies: governmental, 153-66; nongovernmental, 166-79
Respiratory diseases, caused by dust, 194; among miners, 234; *see also* Dust diseases; Phthisis; Silicosis; Tuberculosis
Reznikoff, P., 172
Rhineland, *see under* Prussia

Rhode Island, child labor, 53; population, 60
Rice, G. S., 164, 260, 261, 262
Rille, J. H., K. Ullmann, M. Oppenheim and, 219
Robert, J., 140
Robertshaw, Dr., 200, 201
Rock drillers, 202
Rock dust barriers in coal mines, 230
Rock-dusting in coal mines, 230, 249, 250
Rockefeller Foundation, 254
Rode, C. D., 243
Rogers, C. T. Graham, 62
Romiée, 244
Rosen, George, 9, 225, 233, 238, 239
Rosenfeld, S., 190
Rosenthal-Deussen, E., 204
Rosier, Pilatre de, 18
Ross, A. A., 205
Ross-Smith, Adelaide, 65, 206, 207
—— L. Greenburg, and others, 164
Royal College of Physicians, Social and Preventive Medicine Committee, 103
Royal Commission on Metalliferous Mines and Quarries, 78, 154, 155, 235
Royal Commission on Safety in Coal Mines, 228, 261
Royal Society of England, 10
Royal Society of Medicine, 160
Ruhr District, coal production, 196; regulations for coal mines, 222
Rumania, ceased to use yellow phosphorus matches, 113

Sachs, Otto, 219
Sachs, W., 173
Safety, wage agreements hindered progress of safety measures, 81; collective bargaining on matters affecting, 82; task of state to supervise, in Europe, 84; posters, 151; standard of, in mines, 225; *see also* Accident prevention
Safety Code Correlating Committee of the American Standards Association, 92

Safety committees, 82, 94-99
Safety directors and engineers, 99
Safety education programs, 150-52
Safety First movement, 95, 149, 150-52
Safety in Factories Order, Draft (British), 1927, 95
Safety inspectors, insurance companies' program, 74
Safety lamps for mines, 224, 226, 240 ff.; candle power of various types, 242; loss of candle power in mines, 243
Safety Speakers Bureau, of the National Safety Council, 150
Salmonsen, E. M., 206
Sandblasting, 206
Sandstone, phthisis among workers in, 195
Sanitary Act of 1866 (British), 25
Sanitation and cleanliness, regulations re, 24, 28, 126
Saranac Laboratory, 206
Sauvages, Boisin de, 195
Saxonia, factory inspectors, 46
Saxony, *Gewerbegesetz*, regulation on child labor and health of workers, 41; mine law, 222
Sayers, R. R., 59, 162, 163, 182, 206, 230
Scabies, 216, 218
Scalione, C. C., 143
Scaphander apparatus, 209
Scheffler, C. L., 12
Schlagwetter Commission (Prussia), 247
Schlattmann, 223
Schlockow, I., 197, 198
Schneider, H., 77
Schönlank, Bruno, 172
Schrenk, H. H., 163, 230
—— J. B. Littlefield, F. L. Feicht, and, 164
—— J. W. Ballard, H. J. Oshry, and, 164
—— S. J. Pearce and, 164
Schrötter, H. von, R. Heller, and W. Mager, 113, 210, 212, 214
Schuchard, 41, 86
Schuler, Friedrich, 155

Schwartz, Louis, 162
—— and L. Tulipan, 219
Sciences, progress in, and its influence on industrial hygiene, 180-93; toxicology, 180-82; physiology, 180; statistics, 182-93
Seamen Laws, 57
"Secretage," 115
Self-containing oxygen apparatuses, 144-46, 253
Sendtner, J., 185
Servus apparatus, 146
Sewer workers, 19
Shaw, N. H., 205
Sheafer, H. C., 53
Sherman, Chief Factory Inspector, New York, 62
Shoemakers, age at death, 184
Shop committees, 98
Shower baths, *see* Bathing installations
Sicard, 209
Sickness funds, statistics: reports, 189-93 *passim*; illness excluded from benefits, 192
Sickness insurance, *see* Health insurance; Sickness funds
Siegal, W., 207
Silica dusts, phthisis from exposure to, 201; dusts dangerous, characterized by uncombined crystalline, 203
Silicates, dusts of, 207
Silicic acid, 198
Silicosis, 202 ff.; avoiding, in steel foundries, 36; publications of U.S. Public Health Service on, 162; study of, among rock drillers, blasters, and excavators, 206; compensation for, 208; among South African miners, 237
Silk workers in Lyon, 19
Silver mines, Laurion, 232
Silver nitrate, replaced mercury in production of mirrors, 114
Simson, F. W., and A. Sutherland-Strachan, 203
Simons, Dr., 154
Sinclair, George, 209
Singstad, Ole, 212

INDEX 339

Skin diseases, occupational, 215-19; compensable, 217; hygiene and methods of avoiding, 217
Skins, see Hides
Skragge, Nicolaus, 12
Slate mines, 200
Slaves, Roman, used in mining mercury, vii
Smelting plants, first industrial hygienic installations, 7; exhaustion of vapors in, 8; ventilation, 15
Smith, Adam, 39
Smith, G. W., 131
Smith, J., 100
Smyth, H. F., 141
Snell, E. H., 212, 214, 244, 245
Society of Pottery Workers, 79
Sommerfeld, Theodore, 88, 174, 199
South Africa, study of Miner's phthisis, 141, 202 ff., 235; Institute for Medical Research publications, 203; mine regulation, 236; dust content of air in gold mines, 237; silicosis among miners, 237; study of dust diseases, 282
Southam, A. H., 167
South Carolina, Labor Law: race segregation, 69
South Dakota, population per square mile, 60
South Wales, pulmonary disease in mines, 205
Spaniards introduced work turnover in America, 14
Spanton, W. B., 154
Special problems, 194-219; dust diseases, 194-208; caisson and tunnel work, 208-15; skin diseases, 215-19
Spedding, Carlyle, 246
Spedding, James, 234
Sponge divers, laws concerning, 213
Stahl, G. E., 13
State Health Departments, industrial health units, 66
State ownership of minerals in earth, 220
Statistics, on mortality and morbidity, 182-93
Staub-Oetiker, H., 204
Stauffer, Hans, 217

Steam engine, invention, 16
Steam factories, England, 37
Steam machine, first German, 37
Stegmüller, P., K. Beck and, 160
Stein, Freiherr von, 40
Stern, A. C., 164
Stern, B. J., 99, 110
Steuart, W., 202
Stockhausen, Samuel, 10, 100, 171, 194
Stonecutter's phthisis, see Phthisis
Stone dust, 197
Stratton, Thomas, 196
Strauss, David C., 65
Sugar tube, 130, 132
Suicide, coal vapors used for, 13
Summons, W., 201
Sunderland Society, 224
Surgeons, 100; certifying (examining), 24, 101, 117; "appointed," 120
Süssmilch, Johann Peter, 183
Sutherland, L., 204
Sutherland-Strachan, A., F. W. Simson and, 203
Swiss Federal Council, effort to discuss labor questions with other states, 87
Switzerland, factory act, 49

Tabershaw, J. R., 166
Tailors, sagging stomachs, 4; death from tuberculosis, 184
Talc, effect on lungs, 207
Tanquerel des Planches, L., 140, 171
Technical devices, control, 127-46
Technical inspectors, Germany, 45
Teleky, L., 88, 156, 172, 173, 186, 190, 204, 219
—— H. Katz, and, 171
Temperature, of workroom, 32
Textile industries, law protecting children in, 22; regulations re temperature and humidity, 27; spindles driven by animals and by water power in Saxony, 37
Thackrah, C. T., 19, 52, 195
Theologians, age at death, 184
Thiele, A., 156
Thompson, W. G., 170
Thomson, R. M., 131

Thomson, W., 196
Thorpe, J. W., 133, 228
Tinning of hollow ware, health register required, 120
Tolman, W. H., 55, 169
Toxicology, 180-82
Tracy, Roger S., 53
Trades, mortality, 184 ff.; health rating, 192; *see also* Dangerous trades
Trades Union Congress, 79
Trade unions, membership in U.S., 1836, 51; workers' unions, 75-86; struggle of German, against occupational diseases, 76; English, interested in hygiene and compensation for occupational diseases, 78; different procedures of European and American, 86
Transportation, federal act re interstate, 56
Trasko, V. M., J. J. Bloomfield, and others, 98, 99
Traube, Ludwig, 197
Tremenheere, first mine inspector, 224
Trier, regulation for protection of workers, 42
Trinitrotoluene, protective measures for manufacture of, 101; investigation into toxicology of, 160
Tubercule bacilli, discovery of, 199
Tuberculosis, 199 ff.; caused by dust, 194; *see also* Dust diseases; Phthisis; Respiratory diseases; Silicosis
Tulipan, L., Louis Schwartz and, 219
Tunnel work, 208-15; illness of workers, 211
Tyndall, 140

Ulcers of the legs, 215
Ullmann, K., M. Oppenheim, and J. H. Rille, 219
Unemployment insurance, Great Britain, 51
Union of Factory Workers, 77
Union of the German Accident Insurance Associations, 151
United Automobile Workers, 85, 98
—— C.I.O. Health Institutes, 85

United Mine Workers of America, agreement with State of Washington re local joint safety committees, 82; agreement with U.S. Coal Mines Administrator, 83, 231; Welfare and Retirement Fund, 84 f.
United States, economic development, 51; acts regulating hours and holidays of governmental employees, 56; preventing accidents and disease rests with the states, 57; state labor laws, 58; work committees and safety committees, 98-99; factory physician, 108-11; prohibitive tax on white phosphorus matches, 113; bodies which carry on research and scientific studies in industrial hygiene, 161; books on occupational diseases, 173; periodicals on industrial hygiene, 177; works on toxicology, 181; mines and miners, 228 ff. (*see also* Miners; Mines); mine supervision the task of individual states, 228; coal output, 228; rescue work in mines, 253; fatal coal mine accidents, *tab.*, 259; fatality rates by states, 259, 260; rate of fatal accidents has declined only slightly, 260; qualifications and positions of mine inspectors lower than in Europe, 262; accident mortality rate higher than in Europe, 262; accident fatality rate by industries, *tabs.*, 274, 275; accident fatality rates higher than in Germany, 276
—— Bureau of Labor, *Bulletin*, 54; publications, 164
—— Bureau of Labor Statistics, 271, 273, 275
—— Bureau of Mines, 59, 92, 137, 182; codes, 84; test of dust masks, 141, 142; search for mine rescue apparatus, 145; publications, 163, 230; Technical Bulletin, 205; investigations of health conditions in mines, 205; as a research institution, 229 ff.; staff, 230; has no right to issue rules or orders, 231;

initiated training of men for rescue work, 253; influence, 263
—— Coal Mines Administrator: agreement with United Mine Workers of America, 83, 231
—— Department of Labor, 58; arranged training courses for state factory inspectors, 73
—— Federal Employees' Compensation Act, 74
—— Federal Mine Safety Code, 231
—— Mine Safety Board: re methane in air, 241
—— National Bureau of Standards, 59
—— Public Health Service, 59, 66, 92, 108, 110, 116; publications, 162; Bulletin, 205, 206, 207; investigations of health conditions in mines, 205; Division of Industrial Hygiene, 59
United States Longshoremen and Harbor Workers' Compensation Act, 74
United States Naval Medical Bulletin, 214
United States Steel Corporation, Bureau of Safety, 148
Ursinus, M. Leonardus, 9, 194

Vane, R. J., 268
—— L. I. Dublin, and, 186
Vegetius, F. R., 143
Ventilation, of smelting works, 15; British laws re, 31, 32; artificially humidified air, 32; of rooms in grinding factories, 33; U.S. state laws concerning, 67; developed by industrial science, 282; see also Mine ventilation
Verband der Wiener Genossenschaftskrankenkassen, 189
Vernois, Maxime, 216
Vernon, H. M., 148, 159
Vienna, fire brigade rescue apparatus, 143, 144; workers illnesses with disability caused by lead poisoning, tab., 278
—— Collegium of Physicians, 124
Virchow, R., 197

Virginia, Labor Law, 69
Vital statistics, 182-93; see also Mortality rate
Vitreous enameling, health regulation for workers, 120
Vogt, A., 199

Wade, T. W., 204
Wages, minimum, U.S., 57
Wagner-Murray-Dingell Bill, 80
Wallsend, cause of mine disaster, 247
Walsh-Healey Act, 57
Washington State, agreement with United Mine Workers of America re local joint safety committees, 82
Watelle, T. J., B. Pol and, 210, 211
Water used to decrease dust in air of mines, 226, 235, 236, 249
Watkins-Pitchford, W., 202
Watson, A. W., 189
Watt, A. H., L. G. Irvine, and others, 202
Watt, James, 16
Watt, R. J., 84
Wayne University, classes in health and safety, 85
Weber, H. H., 134, 161
Webster, S. H., 162
Weigels, Christian, Ständebuch, 11
Weinberg, W., 185
Weindl, Casper, 247
Wepfer, F. I., 195
Westergaard, Harold, 186, 189
Weston, H. C., 159
West Virginia, coal output, 228; mine inspectors, 229; accident fatality rates in industries, 274; higher than in Great Britain, 275; than in Germany, 276
Weyl, T., 199, 200, 216
Wheeler, R. V., 228, 243, 249
White, R. Prosser, 218, 219
White lead, struggle against use of, for house painting, 76; manufacture and use of, 88
White lead factory, law regulating, 26, 27; health examination of workers, 120
White lead paint, paints to replace, 114

342 INDEX

Whitney, Jessamine S., 188, 267, 268
Wieck, E. A., 82, 229
Willan, Robert, 215
William and Mary of England, 223
Williams, J., 62
Wilson, S. R., 167
Winslow, C. E. A., 206
Wintgens, E., 156
Wisconsin, population per square mile, 60
Wolf lamp, 240; candle power, 242
Wöhler, 16
Women, working hours, 23; protective measures, 26, 29, 66; provision of seats for, 28, 67; first woman factory inspector, 34; nightwork, 88; susceptibility to particular poisons, 105, 118; exclusion from dangerous work, 116-18, 120, 224; regulations re work in mines, 227; lead poisoning among, 278
Wood, W., 53
Woolen mills, regulations on disposal of waste, 41
Workers, no rules for protection of, in Middle Ages, 5; new class, 16; rising standards: political influence, 17; British, received right to vote, 25; federal authorities in field of labor protection in U.S., 59; cooperation in preventive work, 126-27; safety instruction, 150; health position of American, 268
Workers' unions, *see also* Trade unions
Working day, hygienic, *see* Working hours, shorter
Working days in year, Middle Ages, 5

Working hours, in metal mines, viii; in Middle Ages, 5; need to diminish, 19; for children, 22; for women, 23; U.S. governmental employees, 56; civil aeronautics, 57; shorter, 124-26
Working Men's Party, 51, 52
Workman's compensation, in medieval Germany, viii; in Europe, 48-50; for accidents occurring on way to and from place of employment, 49; value of, for control of occupational diseases, 50; states' provisions for in U.S., 73-74; federal laws, 74
Workmen's Compensation Acts (British): 27, 48, 49; statistics, 270
Workmen's Insurance Association, Germany, 44
Works committees, 94-99
Workshops, first steps in hygiene of, 7-11; overcrowding, 27; sanitary standard for, established, 80
Work turnover, 124
Wright, Carroll D., 54; appointed first U.S. Commissioner of Labor, 58
Wyatt, S., 159

Yaglou, C. P., 182
Yant, W. P., 163

Zangger, Heinrich, 175, 204
Zenker, F. A., 197
Ziemssen, H. von, 53, 169
Zinc oxide, 114
Zitzke, F., 219
—— E. Bering and, 219

Bei Fragen zur Produktsicherheit wenden Sie sich bitte an:
If you have any questions regarding product safety,
please contact:

Walter de Gruyter GmbH
Genthiner Straße 13
10785 Berlin
productsafety@degruyterbrill.com